FOSTERING FLEXIBILITY IN THE ENGINEERING WORK FORCE

Committee on Skill Transferability
in Engineering Labor Markets

Office of Scientific and Engineering Personnel

National Academy Press
Washington, D.C. 1990

NOTICE: The project that is the subject of this report was approved by the Governing Board of the National Research Council, whose members are drawn from the councils of the National Academy of Sciences, the National Academy of Engineering, and the Institute of Medicine.

This report has been reviewed by persons other than the author according to procedures approved by a Report Review Committee consisting of members of the National Academy of Sciences, the National Academy of Engineering, and the Institute of Medicine.

The National Academy of Sciences is a private, nonprofit, self-perpetuating society of distinguished scholars engaged in scientific and engineering research, dedicated to the furtherance of science and technology and to their use for the general welfare. Upon the authority of the charter granted to it by the Congress in 1863, the Academy has a mandate that requires it to advise the federal government on scientific and technical matters. Dr. Frank Press is president of the National Academy of Sciences.

The National Academy of Engineering was established in 1964, under the charter of the National Academy of Sciences, as a parallel organization of outstanding engineers. It is autonomous in its administration and in the selection of its members, sharing with the National Academy of Sciences the responsibility for advising the federal government. The National Academy of Engineering also sponsors engineering programs aimed at meeting national needs, encourages education and research, and recognizes the superior achievements of engineers. Dr. Robert M. White is president of the National Academy of Engineering.

The Institute of Medicine was established in 1970 by the National Academy of Sciences to secure the services of eminent members of appropriate professions in the examination of policy matters pertaining to the health of the public. The Institute acts under the responsibility given to the National Academy of Sciences by its congressional charter to be an adviser to the federal government and, upon its own initiative, to identify issues of medical care, research, and education. Dr. Samuel O. Thier is president of the Institute of Medicine.

The National Research Council was organized by the National Academy of Sciences in 1916 to associate the broad community of science and technology with the Academy's purposes of furthering knowledge and advising the federal government. Functioning in accordance with general policies determined by the Academy, the Council has become the principal operating agency of both the National Academy of Sciences and the National Academy of Engineering in providing services to the government, the public, and the scientific and engineering communities. The Council is administered jointly by both Academies and the Institute of Medicine. Dr. Frank Press and Dr. Robert M. White are chairman and vice chairman, respectively, of the National Research Council.

This material is based on work supported by the National Academy of Engineering's Technology Agenda Program.

Copies available from:
Studies Section
Office of Scientific and Engineering Personnel
National Academy of Sciences
2101 Constitution Avenue, NW—Room GR 402
Washington, DC 20418

Library of Congress Catalog Card Number 90-61001
International Standard Book Number 0-309-04276-3

Printed in the United States of America

COMMITTEE ON SKILL TRANSFERABILITY IN ENGINEERING LABOR MARKETS

NAE Dale R. Corson, Chair
President Emeritus
Cornell University

Robert K. Armstrong
Manager, Professional Staffing
E.I. duPont de Nemours & Co., Inc.

NAE W. Dale Compton
Gilbreth Distinguished Professor of Industrial Engineering
Purdue University

Robert C. Dauffenbach
Director, Bureau of Business & Economic Research
Oklahoma State University

Richard A. Ellis
Director of Manpower Studies
American Association of Engineering Societies

NAE John K. Hulm
Chief Scientist
Westinghouse Electric

Liaison to OSEP Advisory Committee on Studies and Analyses

NAE Lotfi A. Zadeh
Department of Computer Sciences
University of California - Berkeley

Staff Officer

Cheryl B. Leggon

FOREWORD

The pace of development of new technology and the rate at which resulting new products and services enter the marketplace have been steadily accelerating since World War II. Moreover, in this period not only markets but also finance, investment, and industry have become increasingly global in nature. Industrial operations—research, design, development, production, distribution, sales, and service support—even for single products, now take place with less and less regard for national boundaries. In the global economic and industrial complex that is evolving, the engineering enterprise in all nations is constantly challenged to excel in developing and applying new technologies, bringing forth new products and serving new user and societal needs and wants. Thus, in the meetings and forums of the National Academy of Engineering it has been widely recognized that flexibility and adaptability are important qualities for engineers and are destined to be even more important in the future.

Yet, although the positive merits of flexibility and adaptability of engineers are generally regarded as obvious, it is not so clear how these qualities can be assessed and how rapid changes over time in both challenges and responses can be tracked. A 1985 National Research Council report* on engineering education and practice in the United States concluded that the United States did have a flexible, adaptable engineering work force as evidenced by the ability of engineers to transfer among diverse disciplines, engineering functions, and industrial sectors, often in activities well removed from the fields in which they were originally educated. This observed flexibility and adaptability of engineers is often attributed to the solid grounding in basic science, mathematics, and engineering fundamentals received by engineers in their professional education and no doubt this may be the most important contributing factor. But it is not at all clear that sound

---

* National Research Council, *Engineering Education and Practice in the United States*, Washington, D.C.: National Academy Press, 1985.

basic education, particularly limited to only the first professional degree level attained by the majority of engineers, can entirely account for the flexibility and adaptability of engineers experienced in the past, much less be relied on to meet the greater challenges of the future. It is increasingly apparent that a university education, career-long education, personal professional development, and the climate of engineering practice in industry all must contribute to the creation and maintenance of an engineering work force capable of providing technological and industrial leadership for the nation in a highly competitive global economy.

To look at the questions of what might be the changing requirements for the flexibility and adaptability of engineers in the highly competitive global economy and the rapidly changing technologies we face in the future and to examine some of the ways in which these requirements might be met, the NAE asked the National Research Council Office of Scientific and Engineering Personnel to undertake this exploratory study. The findings and conclusions illuminate what can be determined from available data and arrived at by judgments based on past experience as a guide to academic, industry, and government policy. They also indicate some important gaps in our knowledge and understanding of the nature and extent of the future needs for flexibility and adaptability of engineers and of the measures by which such needs may be met.

Alexander H. Flax
Home Secretary
National Academy of Engineering

ACKNOWLEDGMENTS

The Committee on Skill Transferability in Engineering Labor Markets appreciates the assistance that it received from a number of individuals. The National Academy of Engineering (NAE) initiated the establishment of this committee. The committee is most grateful for the many contributions made to the development of this exploratory study by NAE staff: Robert M. White, president; Alexander H. Flax, home secretary; Gerald P. Dinneen, foreign secretary; and Bruce Guile and Samuel Rod, director and associate director, respectively, of the NAE Program Office. Alan Fechter, executive director of the Office of Scientific and Engineering Personnel (OSEP), provided guidance and helpful counsel throughout the course of the study and critical readings of all drafts of the report. Lotfi A. Zadeh, liaison to OSEP's Advisory Committee on Studies and Analyses, contributed critical insight to the committee's deliberations. Cheryl B. Leggon, staff officer, kept the committee abreast of current research, organized the various committee activities—including the workshop convened on September 29, 1989, and wrote the numerous drafts of this report. Michael G. Finn, OSEP director of studies and surveys, provided methodological guidance. Linda S. Dix, reports officer, edited this report; and Linda Simmerson assisted during both the workshop and the production of this report.

Much of the information presented in this report was gathered during the committee-sponsored Workshop on National Needs and Technological Change: Fostering Flexibility in the Engineering Work Force. The committee appreciates the background materials prepared by the following individuals: Pamela H. Atkinson, University of California-Berkeley; Larry M. Blair, Oak Ridge Associated Universities; Robert C. Dauffenbach, Oklahoma State University; Alan Eck, Bureau of Labor Statistics; Michael G. Finn, OSEP; Saul Gorn, University of Pennsylvania; Cheryl B. Leggon, OSEP; J. S. Watson, Oak Ridge National Laboratory; and David Woodall, Idaho National Laboratory. Special acknowledgment is extended to Peter Cannon, president and chief executive officer of Conductus, Inc., for delivering the charge to the workshop and making insightful contributions to the deliberations. The committee thanks the 40 individuals from industry, academe, government, and professional engineering societies who participated in the workshop. Finally, the committee gratefully acknowledges the reviewers, whose comments significantly enhanced the quality of the final report.

We hope that the efforts of the many individuals involved in this exploratory examination of flexibility in the engineering work force will clarify the issues and assist OSEP and NAE in developing their research agenda for the next decade.

Dale R. Corson
Chairman

CONTENTS

EXECUTIVE SUMMARY

Background

Historically, the United States' engineering work force has been able to meet rapidly expanding needs for engineers in such fields as aeronautical engineering, computer engineering, and materials science and in areas driven primarily by national policy, such as defense, energy, and space. During the 1920s, when European universities were the capitals of science, the United States was preeminent by almost any industrial measure—including worker productivity, per capita income, and trade surplus. Now, however, U.S. universities are the capitals of science, and Japan has the trade surplus.[1] Any technology-based advantages held by U.S. firms over foreign firms are likely to be more fleeting in the future not only because new knowledge and technologies developed in the United States are transferred to foreign competitors more rapidly than they were in the past, but also because technology-based advantages will originate in other countries. As industrial competition increases, the product cycle is becoming shorter, while the time needed to complete an engineering degree is becoming longer. This means that now—more than ever before—the demands of industry are more likely to change while a student is enrolled in an engineering program, and the student will need to adapt faster to changes in both technologies and markets; this also means that practicing engineers must remain current throughout their professional careers.

The National Academy of Engineering (NAE) asked the National Research Council's Office of Scientific and Engineering Personnel (OSEP) to conduct a study of what is known and what needs to be known about adaptability among the engineering work force in the United States, as a prerequisite to understanding what policy mechanisms may be needed, and what action might be taken to enhance the adaptability of the U.S. engineering work force. OSEP convened a steering committee of engineers and scientists

[1]Ralph E. Gomory, "From the 'Ladder of Science' to the Product Development Cycle," *Harvard Business Review* 67(6):99-105, November-December 1989.

from industry, academe, and the professional engineering societies to guide its efforts and to prepare a report summarizing its findings and conclusions.

Although adaptability is not directly observable, it can be inferred from mobility patterns among fields, work activities, and sectors of the economy. The committee commissioned special data tabulations and papers reviewing what the major data bases tell us—albeit indirectly—about adaptability. To supplement this quantitative information, case studies were prepared for the fields of chemical engineering, nuclear engineering, and computer science. To assess the extent to which both the quantitative and qualitative knowledge bases reflect reality, the committee invited engineers—and those who educate, train, employ, and study engineers—to an intensive one-day workshop on National Needs and Technological Change: Fostering Flexibility in the Engineering Work Force. The committee's findings and conclusions are based on information developed from these activities.

Findings

The committee agreed that adaptability is not a new issue but that the context in which we examine adaptability is new in two major ways:

1. *Whole complexes of new technologies—for example, information technology, biotechnology, and genetic engineering—require rapid adjustment to their use and commercial exploitation.*

In addition, the committee found that:

2. *The erosion of its technological leadership means that the United States is less able to control the pace at which new technologies and knowledge are disseminated; consequently, the need increases for the United States' economy to respond to new challenges from international scientific, engineering, and economic communities.*

3. *There does not appear to be a lack of mobility among individuals in the engineering work force.*

Analyses of special tabulations of data on persons with engineering degrees identified two groups: one group who are "stayers" in the same engineering field and another group of "switchers" who, once they have switched, continue to switch at relatively high rates either to other engineering fields or to nonengineering fields. Preliminary findings indicate that age does not appear to be related to employment field switching for those with the B.S., M.S., or Ph.D., degrees in engineering. However, the following factors do appear to be related to employment field switching for those with the B.S., or M.S., degree in engineering: number of years since the last degree in engineering was earned, field of study, and primary work activity. For those at the B.S., or M.S., level the data indicate the existence of two career tracks—technical and management; there does not appear to be a dual career track for those with the doctorate in engineering.

4. *In quantitative terms, there does not appear to be a lack of adaptability among the U.S. engineering work force. Extant data are inadequate to assess the qualitative aspects of adaptability.*

5. *Adaptability results as much from socialization as education.*

6. *Adaptability among engineers is enhanced by good grounding not only in the basics—mathematics, statistics, physics, chemistry, and engineering—but also in problem-solving.*

7. *What impedes production of adaptable engineers may not be the engineering curriculum but how that curriculum is delivered.*

8. *Continuing education could enhance adaptability in the engineering work force.*

9. *Taxonomies usually used to collect data on engineers are not adequate to capture the kinds of shifts of specialties and skills occurring in the engineering labor market.*

At best the data give indications of the quantitative aspects of mobility—that is; the numbers of engineers moving from one field to another—but fewer indications of the qualitative aspects of mobility—such as the quality of the work performed by those who move.

Conclusions

- Today, the United States is competing in a challenging arena characterized by unprecedented technological change and rapid transfer of information. Many people believe that successful competition under these conditions requires an even more adaptable engineering work force.
- Current evidence on the adequacy of existing mechanisms to facilitate adaptability is not sufficient to guide the development of policy. Action to improve our knowledge base is required now—before such policy is needed.

If adaptability encompasses the ability to transform new scientific and technological knowledge into product and process applications, then adaptability is facilitated—or impeded—by the way in which engineers and other technical employees are supervised and managed. Further research is needed on the aspects of daily on-the-job activities that facilitate and impede adaptability, including

- the methodology or process aspect of engineering—that is, the ways in which engineers tackle their problems and
- the allocation of project assignments.

Moreover, data on demonstrations and pilot projects—and assessments thereof—undertaken by industry, academe, and the professional associations need to be collected and analyzed to increase our understanding of how to enhance adaptability in the engineering work force.

Both the committee and the workshop participants agreed that there is an urgent need for a forum in which key players from the public and private sectors can meet not only to raise issues but also to devise, implement, evaluate, and track their solutions. Before undertaking any additional activities, the committee suggests that the recommendations from past activities be reviewed to see whether they were implemented and with what outcomes and that a systematic plan be devised to build a solid research base from which to develop additional activities.

1
BACKGROUND

The Academic Advisory Board of the National Academy of Engineering (NAE) identified the adaptability of the engineering work force as an issue with important implications for the United States' ability to exercise technological leadership and compete effectively in international markets. Adaptability has been one of the strengths of the United States' science and engineering work forces. For the purposes of this report, "adaptability" is defined as the ability to transfer a given set of engineering skills among engineering fields and activities, between engineering and nonengineering fields and activities, and among sectors of the economy; transform new scientific and technological knowledge into product and process applications; and seek out and apply ideas from outside sources to the engineering process when needed. As defined, "adaptability" is not only reactive in the sense of being limited in its capacity to adjust to change, but also proactive in the sense of having the flexibility necessary to initiate change.

Historically, the United States' engineering work force has been able to meet rapidly expanding needs for engineers in such fields as aeronautical engineering, computer engineering, and materials science and in areas driven primarily by national policy, such as defense, energy, and space. For example, adaptability facilitated the adjustment to sudden unexpected declines in demand in such areas as nuclear power and petrochemicals: engineers in these areas were able to find jobs elsewhere; unfortunately, the data are sparse on the quality of their performance on these new jobs. Now, however, the United States is in a fundamentally new situation in which competitiveness is qualitatively different from what it was in the past. As our national priorities shift and as our economy becomes more service-oriented, American society must be able to move workers around in response to changes in demand. The question is "Does the arrangement that we now have provide us with the kind of flexibility that will be needed as we look toward the future and the kinds of forces that we will face as a nation and as an engineering community?"[2]

Accordingly, NAE asked the National Research Council's Office of Scientific and Engineering Personnel (OSEP) to conduct a study of what is known and what needs to be known about adaptability among the engineering work force in the United States. The overall objective of this activity is to increase our understanding of adaptability as a way to enhance the United States' capabilities to meet changes in both technology and national needs. Three major outcomes were anticipated:

- a background paper summarizing what is known and what needs to be known about adaptability;
- a one-day workshop to evaluate the state of existing knowledge about adaptability by assessing how well this knowledge informs the major policy issues associated with adaptability, identifying important gaps in this knowledge, and formulating a long-range research agenda that would ultimately reduce these gaps; and
- a proceedings volume based on the workshop discussion.

In response to this request, OSEP convened a steering committee of engineers and scientists from industry, academe, and the professional engineering societies to guide its efforts and to prepare a report summarizing its findings and conclusions.

The committee concurs with the finding from a recent National Research Council report that since the mid-1960s, as the economy has become increasingly international, the United States' economic performance has deteriorated and changes in the international economic environment have narrowed the technological gap between the United States and other industrial economies in many industries.[3] Any technology-based advantages held by U.S. firms and workers over foreign firms and workers are likely to be more fleeting in the future, not only because new knowledge and technologies developed in the United States are transferred to foreign competitors more rapidly than they were in the past, but also because technology-based advantages will originate in other countries. This means that, in the future, the United States will have less control than previously over the pace at which

---

[2]Robert M. White, "Opening Remarks to Workshop on National Needs and Technological Change: Fostering Flexibility in the Engineering Work Force," September 1989.

[3]Richard M. Cyert and David C. Mowery (eds.), *Technology and Employment, Innovation and Growth in the U.S. Economy*, Washington, D.C.: National Academy Press, 1987.

new knowledge and technologies are disseminated; therefore, it will become increasingly important for the U.S. engineering work force to be able to accommodate to rapidly changing employment priorities without significant stress or strain. Technological skill allows a company—or a nation—to be competitive, but no single technological achievement yields a lasting competitive advantage.[4] "New design and manufacturing strategies are perhaps the most interesting ways of turning technology into advantage."[5]

The cycle time for engineering education—on average, four to five years—is longer than the cycle time for many new technologies and new products. This means that the demands of industry are more likely now than in the past to change while a student is enrolled in an engineering program; consequently, students need to adapt faster to changing technologies and markets now than they did in the past. It seems reasonable to conclude that adaptability will become increasingly important as a way for the United States to adjust to changes in technology and demand.

The committee saw its charge as conducting an exploratory assessment of adaptability. Two basic underlying assumptions informed the committee's deliberations:

- An adaptable engineering work force is a valuable national asset because it facilitates adjustments to technological change and shifting national priorities through field switching, industrial mobility, or reallocation of work activities among engineering functions.
- The relationship between adaptability and education and training is strong and complex: the type and level of education and training—both in school and on the job—play important roles in creating an adaptable engineering work force; and the need for adaptability can have important consequences for education and training policies.[6]

A sizable part of the committee's deliberations focused on conceptualizing "adaptability" because, like quality, adaptability is easier to recognize than to define. To a

---

[4]"Competing Beyond Technology," *Harvard Business Review* 67(6):93, November-December 1989.

[5]Kim B. Clark, "What Strategy Can Do for Technology," *Harvard Business Review* 67(6):94-98, November-December 1989.

[6]Pamela H. Atkinson, "The Relevance of Career-Long Education to Creating and Maintaining an Adaptable Work Force," a paper prepared for the Workshop on National Needs and Technological Change: Fostering Flexibility in the Engineering Work Force, September 29, 1989.

certain extent, adaptability is like the archetypal black box: we know that the input is an engineering work force with a particular mix of skills and that the output is an engineering work force with a different mix of skills. We also know that overall market forces have resulted in meeting at least our quantitative needs for engineers in general, and certain kinds of engineers in particular. We do not know about the quality of these adjustments, nor the factors that facilitate and impede them.[7] After considerable discussion, the committee agreed to define "adaptability" as the ability to accommodate smoothly and efficiently to changes in demand for engineering work by applying products, processes, skills, and resources—including knowledge and human resources—in different ways; adaptability also refers to the ability to exploit new technologies effectively and quickly. The committee broadly defined the "engineering work force" to include individuals who meet at least one of the following criteria: earned a degree in engineering; are employed as engineers; or identify themselves as engineers based on their education and work experience. These definitions are consistent with those used in an earlier NRC study on engineering employment characteristics.[8]

Although adaptability is not directly observable, it can be inferred from mobility patterns among fields, work activities, and sectors of the economy. Therefore, the committee examined the following questions:

- What is known about mobility among engineering fields and between engineering and nonengineering fields? How much of this movement is real, and how much is terminological—that is, an artifact of how and why the data on mobility are collected?
- What are the implications of this knowledge for education and training policy?
- Where are the gaps in our knowledge and how can these gaps be closed?

The committee commissioned special data tabulations and papers reviewing what the major data bases tell us—albeit indirectly—about adaptability. To supplement this quantitative

---

[7]Cheryl B. Leggon, "National Needs and Technological Change: A Background Paper," prepared for the Workshop on National Needs and Technological Change: Fostering Flexibility in the Engineering Work Force, September 29, 1989.

[8]Committee on the Education and Utilization of the Engineer, Commission on Engineering and Technical Systems, *Engineering Employment Characteristics*, Washington, D.C.: National Academy Press, 1985a.

information, case studies were prepared on the fields of chemical engineering, nuclear engineering, and computer science.

To assess the extent to which both the quantitative and qualitative knowledge bases reflect the actual experiences of engineers—and those who educate, train, employ, and study engineers—the committee invited representatives from industry, academe, government, and the professional associations to an intensive one-day Workshop on National Needs and Technological Change: Fostering Flexibility in the Engineering Work Force.

The committee's findings and conclusions are based on information developed from these activities.

2
WORKSHOP ON NATIONAL NEEDS AND TECHNOLOGICAL CHANGE: FOSTERING FLEXIBILITY IN THE ENGINEERING WORK FORCE

Because industry is a major employer of engineers, the committee wanted to have industry representation at this workshop. To achieve this goal, a special effort was made to ensure that approximately 20 percent of the 42 workshop attendees came from industry.[9] Workshop participants were asked to use their professional experiences as a basis for addressing the following questions:

- Is adaptability a problem for the United States' engineering work force?
- Is the engineering work force flexible enough?
- In what ways is the engineering work force not flexible enough?
- What kinds of interventions might be necessary right away to foster flexibility in the engineering work force and by which institutions—industry, academe, government, professional societies?

To maximize the opportunity for dialogue, workshop participants spent most of the day in one of three discussion groups: technological change; changing national priorities; and education and training. The *technological change group* focused on how firms deploy their engineering work force to meet changes in demand for engineering skills resulting from changing technologies. This group looked at differences in responses between emerging and declining technologies (defined as technologies experiencing a temporary decline in demand, not technologies going out of existence).[10] The *changing national priorities group* focused on the impact of changes in national priorities on the need for an adaptable engineering work force. For discussion purposes, national priorities included

[9]This total includes staff and the study committee.

[10]Leggon, *op. cit.*

both defense and nondefense priorities. Nondefense issues included the impact of competitiveness and such environmental concerns as the greenhouse effect and global warming. One major issue considered was the impact on adaptability of a reduction in defense spending. The *education and training group* addressed two issues: how to produce flexible, adaptable engineers and how to reduce the degree of mismatch between the competencies required in industry and the competencies of both new and experienced engineers.

At the start of the workshop, participants were given their group assignments. Each group had a committee member to lead the discussion and a rapporteur to record key ideas. At the end of the day, everyone reconvened to hear the group leaders or rapporteurs summarize the groups' major ideas, points of consensus, findings, and conclusions. Although this format was devised as a way to focus on the major aspects of adaptability that the committee identified, the committee realized that each group would discuss all three topics because they seem to be such inextricably intertwined components of "adaptability."

It is noteworthy that despite different starting points and foci, the three groups converged on engineering education—particularly undergraduate and continuing or lifelong education—as a key factor in creating and maintaining an adaptable engineering work force.

3
FINDINGS

Adaptability as an Issue

Interest in adaptability stems from concern over whether there will be enough well-trained people to do engineering jobs in the future—although rapid technological development makes it virtually impossible to foresee what these new jobs will be. The committee found that adaptability is not a new issue: historically, the United States' scientific and engineering work forces have adjusted to shifts in demand. However, the committee did find that the current context for examining adaptability to be different in two major ways:

1. *Whole complexes of new technologies—for example, information technology, biotechnology, and genetic engineering—require rapid adjustment to their use and commercial exploitation.*[11]

Although technological change is not new to engineering, the rate of change in some areas may make it more difficult than ever before to remain current. Engineering has always been among the faster-changing disciplines;[12] this will probably become more pronounced in the future as new technologies are developed and disseminated more rapidly than in the past.

2. *The erosion of its technological leadership means that the United States is less able to control the pace at which new technologies and knowledge are disseminated; consequently, the need increases for the*

---

[11]These sentiments were also expressed by M. P. Manahan, W. B. Ashton, and G. S. Stacey in "Who's on Watch?," *Manufacturing Engineering*, June 1989, p. 54.

[12]Saul Gorn, "Adapting to Computer Science," a paper prepared for the Workshop on National Needs and Technological Change: Fostering Flexibility in the Engineering Work Force, September 29, 1989.

United States' economy to respond to new challenges from international scientific, engineering, and economic communities.

Some of the ways in which supply has adjusted in the past to changes in demand are no longer sufficient. For example, because of demographic changes in the United States, companies can no longer expect to hire substantial numbers of new engineers every year from traditional sources—that is, white males. We must think in terms of, and act on, bolder initiatives, such as tapping more effectively new sources of engineering talent—women and underparticipating minorities (black Americans, Native Americans, Mexican Americans/Chicanos, Puerto Ricans). Moreover, if the United States is to remain competitive internationally, we must enhance our understanding of what facilitates—and impedes—adaptability to maximize our capacity to adjust smoothly and efficiently to changes in knowledge and technology.

Although technological skill qualifies a nation to enter the competitive arena, no single technological achievement yields a lasting competitive advantage.[13] During the 1920s Europe excelled in scientific discovery, while the United States excelled in worker productivity, per capita income, and trade surplus. Now the U.S. excels in scientific discovery, while Japan excels in trade surplus. Product leadership can be built without scientific leadership, if companies excel at design and the management of production. For example, market shares already lost by U.S. consumer electronics and automobile producers can be attributed neither to failures of new science nor to failures of innovation, but rather to failures to make such refinements as customizing a product for more and more market segments, making it more reliable, and getting it to market more cheaply.[14]

Mobility Among the U.S. Engineering Work Force

3. *There does not appear to be a lack of mobility among individuals in the engineering work force.*

---

[13]"Competing Beyond Technology," *Harvard Business Review* 67(6):93, November-December 1989.

[14]Gomory, *op. cit.*

Mobility refers to movement of various types—changes in employer, occupation, and responsibilities. Estimates of mobility rates from 1976 to 1982 indicate that among engineers, almost 22 percent changed employer, 11 percent changed occupation, and 32 percent changed responsibilities. During this period, a larger percentage of engineers changed responsibilities than did physical scientists, biological scientists, and social scientists.[15] The mobility data tell us that some mobility is self-generated by such things as a better job, and some is forced by such things as layoffs or being fired. This only indicates that engineers are employees subject to the vagaries of the economy; it says nothing about the quality of the move—for example, the ability of a vacuum tube engineer to switch to transistors with a minimal reduction in productivity.

Analyses were done on data from three sample-based surveys of scientists and engineers sponsored by the National Science Foundation (NSF).[16] The data were on persons with degrees in engineering and working as engineers in 1980, who were re-surveyed in 1982, 1984, and 1986; this does not include persons with degrees in engineering who were already working in nonscience or nonengineering positions in 1980. Between 1982 and 1986, the data indicate that about 20 percent of the total with B.S., or M.S., degrees in engineering switched either out of or back into engineering employment. These analyses identified two groups: one group who are "stayers" in the same engineering field and another group of "switchers" who, once they have switched, continue to switch at relatively high rates either to other engineering fields or to nonengineering fields.

First, the following characteristics that might differentiate between switchers and stayers were identified: age; number of years since the last degree in engineering was earned; primary work activity; and degree field of study. Then, comparisons of descriptive statistics were made to identify differences between switchers and stayers on these characteristics. Sampling complexities and time constraints precluded tests for statistical

---

[15]Robert C. Dauffenbach and Michael G. Finn, "Evidence of Adaptability in the Labor Market for Engineers: A Review of Recent Studies," a paper prepared for the Workshop on National Needs and Technological Change: Fostering Flexibility in the Engineering Work Force, September 29, 1989.

[16]Data on persons with a doctoral degree in engineering are from the 1987 Survey of Doctoral Scientists and Engineers, a biennial longitudinal survey conducted by the National Research Council's Office of Scientific and Engineering Personnel. Data on experienced work force participants with bachelor's or master's degrees in engineering are from the 1986 National Survey of Natural and Social Scientists and Engineers, conducted by the United States Census Bureau. Data on 1984 and 1985 graduates with bachelor's or master's level degrees in engineering are from the 1986 Survey of Science, Social Science, and Engineering Graduates, conducted by the Institute for Survey Research at Temple University.

significance. The findings consistently indicate that there appear to be some differences between switchers and stayers. The committee found it useful to examine the characteristics of stayers and switchers by degree level—separating those with bachelor's and master's degrees in engineering from those with the doctorate in engineering.

Ph.D.s in Engineering

Among those with the doctorate in engineering, there are no significant differences in average age between the groups of stayers and switchers. Switching does not appear to be related to the number of years since the doctorate in engineering was earned. Switching employment fields does not appear to be strongly related to degree field of study. Switchers are somewhat more likely than stayers to be employed in business, industry, and government and much less likely to be employed in education institutions. As compared to stayers in terms of primary work activity, switchers are much less likely to be involved in teaching and research and development (R&D); about as likely to be involved in R&D management; and more likely to be involved in other management and "operations/other." However, note that managing engineers or R&D management is in the strictest sense not switching out of engineering but pursuing the management track instead of the technical track.

B.S. and M.S. Degrees in Engineering

Age does not appear to be related to employment field switching for those with the B.S., or M.S., degree in engineering: the difference in average age for those in engineering employment versus those in nonengineering employment is only about three years. In general, switchers to nonengineering employment from engineering employment have had their degrees longer than stayers in engineering employment. Among B.S., and M.S., degree holders in engineering, switchers from engineering to nonengineering employment tended to be more experienced than stayers. This is consistent with anecdotal evidence that older engineers—that is, late 40s and older—often have been laid off and replaced by younger engineers who are both cheaper to employ and well-versed in the new "hot" engineering technologies.

B.S., and M.S., switchers are more likely to be employed in business and industry and less likely to be employed in government than stayers; this could be at least partially attributable to government pensions, which make staying appear to government engineers to be a more attractive option than leaving. As compared to stayers, B.S., and M.S., switchers to nonengineering employment fields are more likely to be involved in the primary work activities of R&D management and other management—and therefore to be managers by occupation. According to the data, those with B.S., and M.S., engineering degrees in civil, mechanical, and petroleum engineering have somewhat higher proportions staying employed in the same field as their degrees.

Among all B.S., and M.S., degree holders in engineering, those with degrees in industrial, chemical, and mining engineering have somewhat higher proportions switching into nonengineering employment. One reason for this might be the poor employment opportunities for chemical and mining engineers at the time of the surveys. Among new B.S., and M.S., engineers (i.e., the classes of 1984 and 1985 surveyed in 1986), the fields with the highest percentage of switchers are chemical, industrial, mining, nuclear, and mechanical. The data indicate that earning a nonengineering degree is positively correlated with switching. For example, those staying in the same engineering field or switching back to engineering from nonengineering tend to have quite low percentages earning a nonengineering degree. However, the data do not enable us to determine whether the degree is earned before, after, or during the switch.

4. *In quantitative terms, there does not appear to be a lack of adaptability among the U.S. engineering work force. Extant data are inadequate to assess the qualitative aspects of adaptability.*

If one indicator of adaptability is the correspondence between occupation and education, then the extent of adaptability among scientists and engineers seems quite large.[17] Analysis of data from the Census Bureau's Survey of Income and Program Participation (SIPP) shows that 45 percent of all working engineers did not have an exact match between their detailed field of employment and the detailed field in which they earned their highest degree; and, perhaps even more noteworthy, 20 percent earned a degree in a nonengineering field.[18] Unfortunately, we lack sufficient data to assess the qualitative aspects of adaptability among the U.S. engineering work force.

---

[17]Dauffenbach and Finn, *op. cit.*

[18]*Ibid.*

Education and Training

While exploring the roles of education and training in enhancing adaptability in the engineering work force, the committee focused on how to reduce the degree of mismatch between the competencies required in industry and the competencies of both new and experienced engineers. First, the committee identified the characteristics of an adaptable engineer:

- technical ability
- curiosity
- ability to see interconnections between scientific discovery and commercial product; new technical developments and their potential application(s); and such corporate functions as R&D, design, engineering, and sales.

Then, the committee tried to identify the kinds of education and training that promote the development of these characteristics.

5. *Adaptability results as much from socialization as education.*

Adaptability is fostered by a mind-set—created in school and nurtured in the workplace—that inculcates into engineers their professional responsibility to remain current. Perhaps the most important lesson of undergraduate engineering education is that continuing education is necessary for engineers to be technically current professionals. Academic faculty could illustrate the need for adaptability by continually providing examples of new applications of products, processes, and procedures; they could teach the importance of flexibility by example—that is, by remaining current themselves. Many engineering schools make the curriculum more consistent with the needs of professional practice. Some institutions allow graduation of engineers whose skills are entirely analytic: these engineers are clearly needed—mainly at the doctoral level—to pursue research careers in industry or in academia. However, it becomes a problem when Ph.D.s teaching undergraduate students—most of whom will become practicing engineers—cannot address in their courses the issues and techniques required for successful practicing engineers.

6. *Adaptability among engineers is enhanced by good grounding not only in the basics—mathematics, statistics, physics, chemistry, and engineering—but also in problem-solving.*[19]

Most engineers go through a two-year common core followed by a two-year upper division sequence that continues development of the core in context and focuses on special topics. Within the constraints of a four-year undergraduate engineering education curriculum, more courses in the basics mean fewer specialized courses. Yet, rapid technological development results in the need for more specialized courses. This exemplifies a widely recognized issue in undergraduate engineering education—that is, the conflict between preparing students to be practicing professional engineers and providing a base for lifelong learning in specialties that may not yet exist.[20] Additions to the curriculum are resisted because they entail either eliminating existing requirements, or increasing the total number of courses required for the degree (which means increasing the length of time required to acquire a B.S. degree in engineering)—or both. Therefore, it is not likely that significant quantities of new core material can be added to meet the projected needs of the profession and increase the adaptability of engineers. One way to add information to a curriculum is to develop better ways to systematize the information. Recent developments in computer and information technologies suggest new ways whereby more information can be taught in a given period of time.[21]

Although most U.S. engineering schools educate well in the science underlying engineering, many engineers graduate from college without realizing that at the core of their chosen profession are such matters as designing products and systems of quality that can

---

[19]Michael L. Dertouzos, Richard I. Lester, and Robert M. Solow, *Made in America: Regaining the Productive Edge,* Cambridge, Mass.: The MIT Press, 1989.

[20]Committee on the Education and Utilization of the Engineer, Commission on Engineering and Technical Systems, *Engineering Education and Practice in the United States: Foundations of Our Techno-Economic Future,* Washington, D.C.: National Academy Press, 1985b, p. 68.

[21]J. S. Watson, "Adaptability in Chemical Engineering," a paper prepared for the Workshop on National Needs and Technological Change: Fostering Flexibility in the Engineering Work Force, September 29, 1989.

be manufactured, implemented, and used in a cost-effective way.[22] In the final analysis, however, industry gives the signals: for example, a computer company will prefer to hire a digital systems "hot shot" rather than a broadly educated but unspecialized graduate.[23]

The committee found that

7. *What impedes production of adaptable engineers may not be the engineering curriculum but how that curriculum is delivered.*

Academic disciplines do not reflect actual engineering work activities. As existing fields and new disciplines develop, these disciplinary boundaries will appear to be even more arbitrary.[24] For example, a recent National Research Council study assessing areas of likely growth in opportunities for chemical engineers identified areas that all involve interdisciplinary activities.[25]

Continuing or Lifelong Education and Training

The committee found that

8. *Continuing education could enhance adaptability in the engineering work force.*

Within the next decade, the concept of lifelong education will gain greater prominence in the workplace as workers of all ages and levels of employment will need to be retrained several times throughout their working life to keep pace with the changing demands of the work environment.[26]

---

[22]Simon Ramo, "National Security and Our Technology Edge," *Harvard Business Review* 67(6):115-120, November-December 1989.

[23]This point was provided by a reviewer of this report.

[24]Although Watson (*op. cit.*) makes this point about undergraduate *chemical* engineering education in particular, it applies to undergraduate engineering in general.

[25]Committee on Chemical Engineering Frontiers, *Frontiers in Chemical Engineering: Research Needs and Opportunities*, Washington, D.C.: National Academy Press, 1988.

[26]Population Reference Bureau, Inc., *America in the 21st Century, A Demographic Overview*, May 1989, p. 19.

Engineering is a profession whose success is measured by its solved problems. Consequently, not only must all engineers be educated and trained throughout their careers, they must also acquire an understanding of the problems in the discipline to which their work is being applied.[27]

Although some workshop participants suggested the need to educate for adaptability, we do not know how to do it. Nevertheless, engineering undergraduate education can impart to students a sense of professional responsibility to keep current throughout their engineering careers. Not only must schools socialize for adaptability, companies must manage for it. Companies should set lifelong education and training as an expectation for employment and then facilitate the realization of this expectation by doing such things as building into product plans and work schedules time for continuing education and training. Further, companies should examine their policies and practices to determine whether the ways in which they do business encourage—or discourage—adaptability in their engineering work force. Do early retirement policies and practices convey the message that industry wants to send to its engineering work force—that is, that their careers are going to peak at age 50? The cycle time for engineering education—that is, on average 4-5 years—is longer than the cycle time for many new technologies and new products; if industry asks its employees to be ready to fill future needs without forecasting what those needs are, industry has not adequately addressed the issue of adaptability. Companies should articulate to their work force where the company sees itself going in the next 10 years in order for employees to assess where their talents and skills will be needed.[28]

Companies have problems with engineers becoming obsolete. Relying on in-house training and guidance is problematic when a company is not sure that it knows which of its senior technical people are technically current. It is difficult not only to identify engineers who do not remain current, but also to motivate those so identified to become current. Obsolescence is a problem—even for newly trained engineers—in fields in which the technology is rapidly changing. Not surprisingly, adaptability is less a problem for new engineering graduates than for more experienced engineers. New engineering graduates have good basic skills and are motivated to continue their education—they want to take

---

[27]Gorn, *op. cit.*

[28]The number of years given is 10 because, depending on the discipline, obsolescence can begin to overtake engineers as early as 10 years after graduation, according to the National Research Council report *Engineering Employment Characteristics*.

courses to increase their understanding of their field and often ask potential employers about educational opportunities; anecdotal evidence indicates that this trend seems more pronounced possibly because today's young engineers are less likely than in the past to expect to work for the same company throughout their careers.

Much of our interest in adaptability stems from concern about whether there will be enough well-trained people to do engineering jobs in the future. Actually, the concern will be more over the readiness of the work force, than its actual size. Between 1989 and the turn of the century, the rate of growth of the U.S. labor force will steadily decline; yet, the number of workers in the labor force is projected to be more than 140 million by the year 2000 (as compared with the 120 million in 1989).[29] Moreover, the composition of the U.S. labor force is changing: between 1985 and 2000, U.S. minorities and immigrants to the United States will account for approximately 42 percent of the growth of the U.S. labor force; white males will account for 16 percent.[30] Because of these changes, companies can no longer expect to hire an adequate number of new engineers from traditional sources—that is, white males. This leaves companies with two choices: either lure talented professionals from their competitors or upgrade their own professionals; of these, the latter has the greatest long-term potential for achieving national economic goals.[31] Although the private sector is expected to forecast future demand and then act on that forecast, the largest single uncertainty is the supply of people.[32] People do not have a problem making these shifts if there is both an economic force to compel them to do it (push) and a shortage of people in the new position (pull). Companies are concerned with moving people from current jobs into new positions but generally lack specific, standardized ways of bringing new skills to those people, and them to the new jobs. The issue is not whether an employer can move an engineer, but whether that engineer can handle the new job.

---

[29]Population Reference Bureau, Inc., *op. cit.*

[30]*Workforce 2000: Work and Workers for the Twenty-First Century,* William B. Johnston, project director; Arnold E. Packer, co-project director; with contributions by Matthew E. Jaffe, et al., Indianapolis, Indiana: Hudson Institute, 1987.

[31]Atkinson, *op. cit.*, p. 3.

[32]From Peter Cannon's charge to the Workshop on National Needs and Technological Change: Fostering Flexibility in the Engineering Work Force, September 29, 1989.

Continued investment in on-the-job training is needed—especially in situations where engineers have highly specialized training and the demand for that training evaporates. According to a recent study, computer-based tools continue to change the practice of engineering dramatically and challenge the engineer's ability to remain current; since it is impossible to predict what these changes will be and where they will take place, achieving technical currency within a business is management's responsibility.[33] Regardless of size, companies "must encourage engineers to keep up-to-date and know what is going on in the outside world so that when the time is right engineers can pull new technology rather than oppose the unfamiliar being pushed at them."[34] This entails commitment not only of time from the engineer, but also of both time and money from the company. Attitudes of line managers are critical in this regard: line managers are willing to give their engineers released time for in-house courses with immediate, short-range application; they seem less willing to do so for courses with less immediate, longer-range application. Although many companies consider education and training to be as crucial to their success as research and development, education funds are often the first cut and the last reinstated when those companies experience economic difficulty.[35] Many engineers do not work in companies with large in-house education and training facilities; therefore, at the present time, their only alternative for continuing education is to return to academe.

When experienced engineers return to academe, they find that the content is not geared to the motivation and learning style of older, experienced engineers but rather to those of the young person training to become an engineer. Few colleges and universities provide instruction through their engineering departments—rather than extension divisions—without in-class examinations; these examinations can be difficult for individuals who have been away for a long time from both the classroom and the student role. This is interpreted as an indication of lack of responsiveness on the part of academe to the needs of the experienced engineer and to industry, which wants high-quality, need-specific, well-produced programs designed for adult learners.[36] Despite their concern about the need for continuing education of engineers in industry, many engineering schools

---

[33]Committee on the Education and Utilization of the Engineer, 1985b, p. 32.

[34]Gomory, *op. cit.*, p. 103.

[35]Atkinson, *op. cit.*, p. 6.

[36]Atkinson, *op. cit.*, p. 10.

view this as industry's problem—not academe's; consequently these schools choose not to be involved. Some academics view current training programs as being more concerned with fixing immediate problems than with preventing future ones; although they decry the lack of structure, quality control, and planning of these programs, they refuse to be directly involved in remedying the situation.

- *Continuing professional development is needed not only for engineers in industry, but also for engineers in academe.*

Because most undergraduate students go into engineering practice—rather than research or academe—engineering professors must be able to address in their courses the issues and techniques required for successful careers as practicing engineers. This means that undergraduate engineering professors should be aware of current engineering practice—particularly in industry. This awareness can be maintained through attending meetings, workshops, seminars, and short courses sponsored by professional engineering societies. One frequently cited example is the Faculty Professional Development Program sponsored by the American Society for Engineering Education (ASEE). This program "is uniquely designed to maximize the inclusion of recent technological developments into the ever-changing undergraduate curriculum. The ultimate goal is to enhance the teaching capabilities of faculty members so they can return to their campuses with new material enabling them to teach with greater knowledge and effectiveness."[37] Senior faculty and corporate staff who are leaders in their disciplines teach the courses, which focus on topics that belong in a modern undergraduate or first-year graduate program rather than emphasizing only the latest research techniques and developments. Since its inception in 1986, this program has been attended by approximately 350 professors of all ranks. According to an ASEE survey, 95 percent of professors who attended this program in 1989 rated it as excellent or very good. Perhaps accreditation organizations such as the Accreditation Board for Engineering and Technology (ABET) could make continuing professional development of faculty a criterion for accreditation.

---

[37]ASEE Faculty Professional Development Program brochure, 1990.

- *An ongoing exchange between academe and industry is needed to better target education and training to meet the needs of industry as well as those of the individual engineer—both new and experienced.*

To facilitate the transition from school into the engineering work force, both faculty and students would benefit from visiting industry to see how it applies processes and techniques taught in the classroom; industry could develop and implement special courses for faculty concerning these applications. Engineers from industry would benefit from visiting colleges and universities to see how engineering is currently being taught so that industry could make better-informed decisions as to what equipment to donate to academe and how best to instruct faculty to use it. To facilitate the participation in such an exchange of engineering educators from smaller and less well known departments or institutions, an organization like ASEE could play a broker role.

Data and Knowledge Bases

9. *Taxonomies usually used to collect data on engineers are not adequate to capture the kinds of shifts of specialties and skills occurring in the engineering labor market.*

For example, the SIPP data provide practically the only evidence available on the correspondence between education and occupation in the U.S. labor force. This evidence is highly aggregated and subject to large sample variation because of the relatively small sample base. The Occupational Employment Statistics Survey (OES) obtains only wage and salary employment data about employees; because data are obtained from employers, the results count jobs, not individuals.[38] The absence of demographic data limits the usefulness of the OES survey in examining the adaptability of engineers. One limitation of the Current Population Survey (CPS) data is that they understate entrants because the CPS excludes individuals who change residences between surveys. Since many individuals are

---

[38]Alan Eck, "Adaptability of the Engineering Work Force: Information Available from the Bureau of Labor Statistics," a paper prepared for the Workshop on National Needs and Technological Change: Fostering Flexibility in the Engineering Work Force, September 29, 1989.

likely to move after completing school, they are not counted in the second survey, even though they are likely to be working as engineers.

It is difficult to determine how much of the movement is real and how much is terminological—that is, due to different survey definitions of engineers in general and subcategories of engineers in particular, as well as to people reclassifying themselves without really changing their activities. Suggestions to remedy this include standardizing data categories or classifications and devising cross-references or translations among data bases. Although the data indicate a fairly high degree of adaptability among engineers—insofar as more than 50 percent of those with bachelor's and higher degrees in engineering work outside science and engineering—they tell us nothing about the extent to which these individuals are both willing and able to return to science and engineering careers.

One major gap in our knowledge at present is the lack of data on the quality of the adaptation of engineers who change fields or functions. Another major gap is the lack of data on those engineers who stay in their field or function but are adaptable insofar as they make "incremental improvements" in products, technology, or process.

Research is needed to find out how the success of the firm is linked to how it uses its engineering work force. The subjects of this research should include not only the engineering work force—traditionally line and bench engineers and technicians—but also project and group leaders, upper-level management, and chief executive officers.[39] Systematic research needs to be conducted to identify the link between continuing education and adaptability—including the relationships between the success of the firm, how it develops and uses its engineering work force, and the amount of money it spends on various types of continuing education and training.

Because so little is known about the costs and effectiveness of programs designed to upgrade engineering skills, pilot projects and demonstrations should be developed and put in operation to increase the stock of knowledge. Moreover, mechanisms should be devised and implemented to collect, evaluate, and disseminate information on these pilot projects and demonstrations. If implemented, these suggestions could enhance the systematic collection, evaluation, and dissemination of information; facilitate the communication of successful interventions; and perhaps encourage their implementation elsewhere.

---

[39] Atkinson, *op. cit.*, p. 1.

Unless there is a way to make it clear to all the key institutional players—industry, government, academe, and the engineering professional societies—that having a flexible engineering work force benefits all of us, adaptability will be seen as somebody else's issue, and some doable changes will not be made. Since the process of change is influenced by top-level endorsement, any initiatives proposed to enhance adaptability among the engineering work force should be endorsed on a national level by an organization that commands the attention and respect of the engineering community. Therefore, it seems warranted to begin planning the kinds of activities—for example, forums, symposia, action-oriented, problem-solving sessions—that bring together these key players to initiate ways to enhance our knowledge as a prerequisite to devising, implementing, and monitoring policy to enhance adaptability among the United States' engineering work force.

4
CONCLUSIONS

The United States faces several national problems—such as unprecedented technological change and rapid transfer of information—that many people believe are approaching crisis levels and whose solutions will require an even more adaptable engineering work force. Evidence on the adequacy of existing mechanisms to facilitate adaptability is not sufficient to guide the development of policy.

Emphasizing mobility as a key aspect of adaptability minimizes the benefits derived by society from engineers who contribute—not by moving and developing new products—but by staying in one place and making incremental improvements in existing products and processes. In the context of nonmobility, adaptability is manifest by being open to new ideas about and new techniques for improving existing products and their production. This "cyclic development" of successive iterations is a competitive process carried out by product engineers to refine the product, customize it for more and more consumer segments, make it more reliable, or get it to market more cheaply.[40]

If the definition of "adaptability" encompasses the ability to transform new scientific and technological knowledge into product and process applications, then adaptability improves a nation's competitive advantage in the global marketplace not only by enhancing the production of more engineers but also by facilitating the production of more versatile engineers who constantly seek improvements in products and processes. Adaptability is facilitated—or impeded—by the way in which engineers are supervised and managed. Further research is needed on the aspects of daily on-the-job activities that facilitate and impede adaptability, including the ways in which engineers solve problems and the allocation of project assignments. Moreover, data on demonstrations and pilot projects undertaken by industry, academe, and the professional engineering associations need to be

---

[40]Ralph E. Gomory, *op. cit.*

collected and analyzed to increase our understanding of how to enhance adaptability in the engineering work force.

The issues of education and training are always with us: scores of committees have produced recommendations for education reform at all levels. One of the most widely quoted was *A Nation at Risk,* which focused on grades K-12.[41] One of the most thorough treatments of engineering education is the Haddad report, which includes separate volumes on undergraduate engineering education, engineering graduate education and research, and continuing or lifelong engineering education.[42] Before embarking on additional initiatives, it would be beneficial to find out whether recommendations from previous studies were implemented, and with what results.

In the committee's view, by exploring adaptability now, the NAE is on a track that is both timely and necessary.

---

[41]National Commission on Excellence in Education, *A Nation at Risk: The Imperative for Educational Reform*, a report to the nation and the Secretary of Education, Washington, D.C.: The Commission, 1983.

[42]Committee on the Education and Utilization of the Engineer, 1985b. This report was prepared under the guidance of the Committee on the Education and Utilization of the Engineer, chaired by Jerrier A. Haddad.

5
NEXT STEPS

Some of the participants described the workshop as catalytic: it not only raised important issues, but also generated enthusiasm and commitment from some participants to address these issues effectively. Both the committee and the workshop participants agreed that there is an urgent need for a forum in which key players from the public and private sectors can meet not only to raise issues, but also to devise, implement, evaluate, and track their solutions.

Specifically, the following action items were suggested:

- Devise and implement a follow-up to this workshop; this could be the prototype for tracking the results or recommendations from other activities—studies, workshops, symposia—targeted to upgrade the skills of the engineering work force.
- Convene a meeting of appropriate representatives from industry, academe, government, and professional societies to plan and implement a demonstration project of the academe-industry exchange discussed during the workshop.
- Before undertaking any additional activities
  - review the recommendations from past studies and task forces, such as those of the NRC Committee on the Education and Utilization of the Engineer (the Haddad committee) to see whether they were implemented and with what outcomes and
  - devise a systematic plan to build a solid research base from which to develop additional activities. This research base should include data on the aspects of daily on-the-job activities that facilitate and impede adaptability—such as the ways in which engineers are managed and the determinants and impacts of career-long education.

COMMISSIONED PAPERS

NATIONAL NEEDS AND TECHNOLOGICAL CHANGE: A BACKGROUND PAPER

Cheryl B. Leggon
National Research Council, OSEP

The purpose of this paper is to provide general information to individuals prior to their participation in the Workshop on National Needs and Technological Change: Fostering Flexibility in the Engineering Work Force, to be convened on September 29, 1989, by the Committee on Skill Transferability in Engineering Labor Markets. To maximize the benefits from an intensive one-day workshop, the committee agreed that workshop participants should be briefed beforehand on the Committee's deliberations so that the workshop could build on them rather than repeat them. Workshop participants will identify the major policy issues concerning adaptability of the engineering work force and evaluate the state of the knowledge base that informs these issues, enabling the study committee to outline a long-range action agenda that focuses on filling in the gaps in knowledge informing the major policy issues associated with adaptability.

Purpose and Structure of the Workshop

The workshop seeks answers to several questions: Is there a problem getting people to fill engineering jobs? If so, what is the magnitude of that problem? How do workshop participants manage the problem? What are the opportunities—as well as the problems? Are there actions that should be taken? Are there any danger signals about which we should be concerned? What research is needed to address these issues?

To maximize the opportunity for dialogue, workshop participants will spend most of the day in one of three groups: Changing National Priorities; Technological Change; Education and Training. These groups were devised as a way to cover each of the major dimensions that the committee identified to assess adaptability. Although each group will focus on one of the major dimensions, it will also discuss the others. For example, the national priorities group's discussion will undoubtedly include technological change and education and training. For each group, the committee identified specific items for discussion.

Changing National Priorities

This group will consider the impact of changes in national priorities on the need for an adaptable engineering work force. For example, passage of the Atomic Energy Act of 1954 established the early development of commercial nuclear power as a national objective, which enhanced the continued development of commercial nuclear power. Other issues the group will consider include:

- how industry adapts to changing patterns of demand—specifically, the work force practices that they believe enhance their ability to respond to environmental changes and
- impact of federal immigration policies on the adaptability of the U.S. engineering work force.

Technological Change

One major issue will concern emerging and declining technologies. Engineering has always been among the faster changing disciplines because

> a solved engineering problem turns into a standard maintenance technique in an action-oriented discipline that is no longer research engineering (Gorn, 1989).

One example of this is the engineering application of x-rays and electro-magnetics to medicine, resulting in radiology. Gorn adds computer engineering as an emerging profession characterized by rapid technological development—"five generations of machines to one generation of people." One consequence of this is that computer engineers must have continuing education—either academically or on the job—to remain current. Similarly, chemical engineering can be viewed as either the youngest of the major traditional engineering disciplines or as the oldest of the major new disciplines (Watson, 1989). Declining technologies are those experiencing a temporary decline in demand, not technologies going out of existence. Nuclear engineering can be considered both an emerging and a declining technology: it can be considered an emerging technology because it is only about 30 years old; it can be considered a declining technology insofar as negative public perception of nuclear power led to a steady decline in enrollment in undergraduate nuclear engineering programs—although, contrary to popular belief, the demand for nuclear engineers was strong during the past 20 years. Accordingly, Woodall's case study of nuclear engineering may provide insight on how to deal with both emerging and declining technologies. Woodall points out that the first nuclear engineers were trained as physicists, chemists, and mechanical or chemical engineers; and even today—because nuclear engineering is a relatively young field—many senior faculty members of nuclear engineering departments have degrees from fields other than nuclear engineering.[1]

Federal policy actions vary enormously depending upon which areas are perceived to be declining and which are perceived to be emerging. With declining technologies, the issue is not only that of adapting the labor force in new directions, but also of identifying the opportunities. For example, in the relatively new area of biotechnology, biotech engineers are less likely to come from engineering than from biology and chemistry. What kind of training is required for them to make the transition? Who is going to pay for it? Examining how both emerging and declining fields view their adaptability problems may enable one to uncover combinations of declining and emerging interests which, if not addressed in advance, create blockages or problems and exacerbate existing barriers to adaptability; conversely, one might also discover avenues of opportunities for the engineering work force.

Thus, basic questions that this group will discuss include the following:

- How in the future will we deal with the new technologies? How are we going to provide the people to use and develop these technologies?
- How do firms adapt to changing technologies?

Education and Training

For this workshop, the terms "education" and "training" will be used interchangeably—although for some people there are significant distinctions between the

---

[1]See Larry Blair, *Education Fields of Early Ph.D. Nuclear Engineers*, Appendix A of this paper.

two terms. This group will look at the role of education and training—at the undergraduate, graduate, and continuing or lifelong levels—in meeting the changing needs of our engineering work force. Education and training includes not only that provided by formal courses of instruction, but also that provided by industry in-house. Among the issues that the group will consider are the following:

- **Broad versus narrower undergraduate engineering programs in promoting adaptability:** Watson (1989) contends that the key to increasing the adaptability of chemical engineers for work in other fields is to include broader and additional training in the chemical engineering curriculum, while Woodall (1989) points out that the nuclear engineering curriculum could be used as a model for the development of an inherently flexible engineering training program. Are there do-able modifications/additions/changes that could be made to current undergraduate engineering programs and tracks that might improve a young engineer's ability to adapt to the changing professional demands of the work place and increase his/her worth in a fast-changing, technologically sophisticated marketplace?
- **Effectiveness of continuing education in promoting adaptability:** Because engineering is a profession whose success is measured by its solved problems, all engineers must continue to be educated throughout their careers and must acquire an understanding of the problems in the discipline to which their work is being applied (Gorn, 1989). However, some academics see current training programs nationwide as being more concerned with fixing immediate problems than preventing future ones and lament these programs' lack of structure, quality control, and program planning; paradoxically, they refuse to be directly involved in improving the situation. Atkinson (1989) contends that technical training and updating of engineers is becoming an important priority on the national engineering agenda because

  > the company that employs the most up-to-date technical work force is the company that is able to use technology to improve its market advantage and its competitive position.

  Changing demographics of the U.S. work force preclude companies from hiring significant numbers of new engineers every year. Therefore, companies have two alternatives: lure talented professionals from their competitors or retrain/upgrade their own professionals. Atkinson believes that the latter strategy has the greatest long-term potential for achieving national economic goals. Industry continues to provide and pay for education for its employees because it recognizes the role that continuing education has in the company's success. Nevertheless, education funds are the first cut when times become difficult, and the last reinstated when they improve.

- **Role of professional societies in facilitating adaptability:** The Institute of Electrical & Electronics Engineers (IEEE) is working with groups of experts to create self-administered tests and questionnaires that will show what field-specific knowledge elements would be required to move into certain new areas (e.g., moving from magnetics into fiber optics).[2]

Adaptability Matrix

To help conceptualize these issues in a manageable way, the committee developed an adaptability matrix in which the rows represent three major perspectives from which to examine adaptability, and the columns are specific items to be included in each examination:

	Problems	Opportunities	Data
Changes in National Priorities			
Technological Change			
Education and Training			

Terminology

To establish a common universe of discourse that will facilitate communication among workshop participants, this section provides brief, basic definitions of the terms

[2]A copy of the IEEE self-assessment is in Appendix A of Atkinson's paper.

used to describe the workshop. For the purposes of this workshop, national priorities refer to both defense and nondefense priorities. Examples of defense priorities include such weapons as the Bigeye binary bomb and the so-called "Star Wars" defense initiative. Examples of nondefense priorities include environmental concerns such as the "greenhouse effect," the depletion of the ozone layer; energy issues; and competitiveness.

Technological change has been a central component of U.S. economic growth: innovations in products and processes resulted in the creation of new industries and the transformation of older ones (Cyert and Mowery, 1987). Technological advance plays a central role both in changing the environment of competition and in providing firms with a capability to excel in their products and processes.[3]

Adaptability is defined as the ability to transfer engineering skills among engineering subfields and to transform scientific and technological knowledge into product and process applications; this includes applying products, processes, and skills in new ways. However, examining the engineering work force, we understand neither the adjustment process itself nor the factors that facilitate and impede it. The engineering work force is broadly defined to include individuals who earned degrees in engineering, or are employed as engineers, or are self-identified as engineers, based on their education and work experience. This definition includes engineers in management, finance, and public policy.

Background

Contemporary American society is characterized by rapidly changing technology and complex international problems including national security and international competitiveness. The United States is now in a fundamentally new situation in which competitiveness is qualitatively different from what it was in the past. Within the last 30 years, the economy has become increasingly international in character. Furthermore, since the mid-1960s, the United States' economic performance has deteriorated and changes in the international economic environment have narrowed the technological gap between the United States and other industrial economies (Cyert and Mowery, 1987). The more rapid

---

[3]Robert M. White, in the preface to *The Technological Dimensions of International Competitiveness*, a report to the Council of the National Academy of Engineering, Washington, D.C.: National Academy Press, 1988.

rates of international technology transfer characteristic of the modern economic environment mean that the knowledge forming the basis for commercial innovations need not be domestic in origin. Because new knowledge and technologies developed in the United States are transferred to foreign competitors more rapidly than they were in the past, any technology-based advantages held by U.S. firms and workers over foreign firms and workers are likely to be more fleeting in the future (Cyert and Mowery, 1987).

To benefit from technological change within the economy, workers must be able to move from sectors of declining labor demand to those in which employment opportunities are expanding. It seems reasonable, therefore, to conclude that adaptability will become increasingly important over time because it provides a way for the United States to respond to changes in demand for specific kinds of engineering skills. As a prelude to the workshop, it might be useful to examine what we know about the engineering work force in general, and its adaptability in particular.

Selected Characteristics of Engineers[4]

One major source of information is the Current Population Survey (CPS), a survey of approximately 55,000 households conducted monthly by the Bureau of the Census for the Bureau of Labor Statistics and providing information on industry and occupation of employment, age, sex, and education. The characteristics of engineers are very similar to those of all professional workers with two notable exceptions: engineering has a smaller proportion of part-time and female workers and tends to be more stable than other occupations. According to occupational tenure data—which measure the length of time individuals have done the kind of work they are now doing, while working for either their current or any previous employer—the median years of tenure in their current occupation was 10.5 for engineers, 9.6 for all professional workers, but only 6.6 years for all employed workers. Another indication of occupational stability is the proportion of workers with 20 or more years tenure in the occupation: 28.2 percent of engineers have 20 or more years as engineers as compared to 20 percent for all professional workers and 14.6 percent for all workers.

Demand for engineers has increased by 83 percent during the past 25 years. The only significant deviation occurred between 1968 and 1973 as a result of decreasing

---

[4]This section summarizes the paper prepared for this workshop by Alan Eck, included in this volume.

involvement in Vietnam and the space program. Demand for additional engineers results from growth and the need to replace workers who leave the occupation; growth is the easier component to identify.

Merged data—which provide a composite description of movements into, out of, and between occupations over a 1-year period—show that relatively few engineers leave from one year to the next. Separation rates are around 6 percent for the most technical groups of engineers—aerospace, civil, electrical and electronic, and mechanical; this is one-half the separation rate for industrial engineers. Merged data can also provide information about entrants (i.e., individuals who entered the occupation to fill newly created jobs as well as to replace engineers who left); however, such data tend to understate the number of entrants because the data cannot identify entrants who recently completed school. Gross flow data indicate about 8 percent of engineers leave engineering from one year to the next: some leave permanently (to become managers or retire), while others leave temporarily to work in another occupation or stop working. The temporary movements are the most difficult to quantify because they are the most affected by market conditions. The data indicate that if more engineers are needed, labor market adjustments are made; but what about the quality of these adjustments?

Quality of Labor Market Adjustments[5]

One measure of adaptability is the willingness and ability of individuals trained in one field of study to work in alternative occupations. Using data from the Survey of Income and Program Participation (SIPP), Dauffenbach found that among the 20,000 observations of employed engineers, about 55 percent had an exact match between their detailed employment field and the detailed degree field of their highest degree earned. He found that 80 percent of all working engineers had engineering degrees, though not necessarily an exact match. However, slightly more than half of those with engineering degrees were employed in jobs other than science and engineering jobs.

Dauffenbach's analysis of SIPP data viewed the prevalence of non-exact-correspondence between occupation and education as evidence of flexibility in the science and engineering labor market. He hypothesized that one negative consequence of such flexibility occurs in the form of diminished productivity, which should show up

---

[5]This discussion summarizes a review of recent studies on evidence of adaptability in the labor market for engineers by Robert C. Dauffenbach and Michael G. Finn, included in this volume.

systematically in salaries. The results from Dauffenbach's study indicate clearly that persons without engineering degrees earn less than those with engineering degrees when the job is engineering. The one exception appears to be the math and computer science degree recipients, who seem to show as much adaptability as engineering degree recipients at least for jobs in the broad categories of engineering, mathematical sciences, computer science, and physical science. Several studies looking at the earnings of engineers in nonengineering jobs found that persons with engineering degrees are very adaptable in the following sense: they do not earn less when moving from engineering to nonengineering jobs.

<u>**Who stays in engineering and who switches to nonengineering fields?**</u> To find out what the data tell us about which people with degrees in engineering stay in engineering and which switch to nonengineering fields, special data tabulations and analyses were done by Larry Blair of Oak Ridge Associated Universities. The data was on persons with degrees in engineering and working as engineers in 1980, who were resurveyed in 1982, 1984, and 1986; this does NOT include the many persons with degrees in engineering who were already switchers to nonscience or nonengineering positions in 1980. The data are from three sample-based surveys of scientists and engineers sponsored by the National Science Foundation.[6] One caveat: this is a preliminary exploration of the data and is not intended to be definitive.

We found it best to examine the characteristics of "stayers" and "switchers" by degree level: separating those with B.S., and M.S., degrees in engineering from those with the Ph.D.

According to Blair's tabulations, during 1982 and 1986, there was considerable switching in general to nonengineering employment and back to engineering employment—with the "switchers" out of engineering approximately balanced by the "return switchers" to engineering. There also appears to be two groups: "stayers" who remain in the same engineering field and "switchers" who once they have switched continue to switch at relatively high rates to either other engineering fields or other nonengineering fields

[6]Data on persons with a doctorate degree in engineering are from the *1987 Survey of Doctoral Scientists and Engineers*, a biennial longitudinal survey conducted by the National Research Council's Office of Scientific and Engineering Personnel. Data on experienced work force persons with bachelor's or master's degrees in engineering are from the *1986 National Survey of Natural and Social Scientists and Engineers*, conducted by the United States Census Bureau. Data on 1984 and 1985 graduates with bachelor's or master's level degrees in engineering are from the *1986 Survey of Science, Social Science, and Engineering Graduates*, conducted by the Institute for Survey Research at Temple University.

(Figure 1). Age does not appear to be related to employment field-switching. In general, "switchers" had their B.S., or M.S., degrees longer than "stayers." "Switchers" tended to be more experienced than "stayers;" however, "switchers" within engineering employment fields tended to be less experienced than "stayers." B.S., and M.S., "switchers" are more likely to be employed in business and industry and less likely to be employed in government than "stayers." B.S., and M.S., "switchers" to nonengineering employment fields are more likely to be managers by occupation.

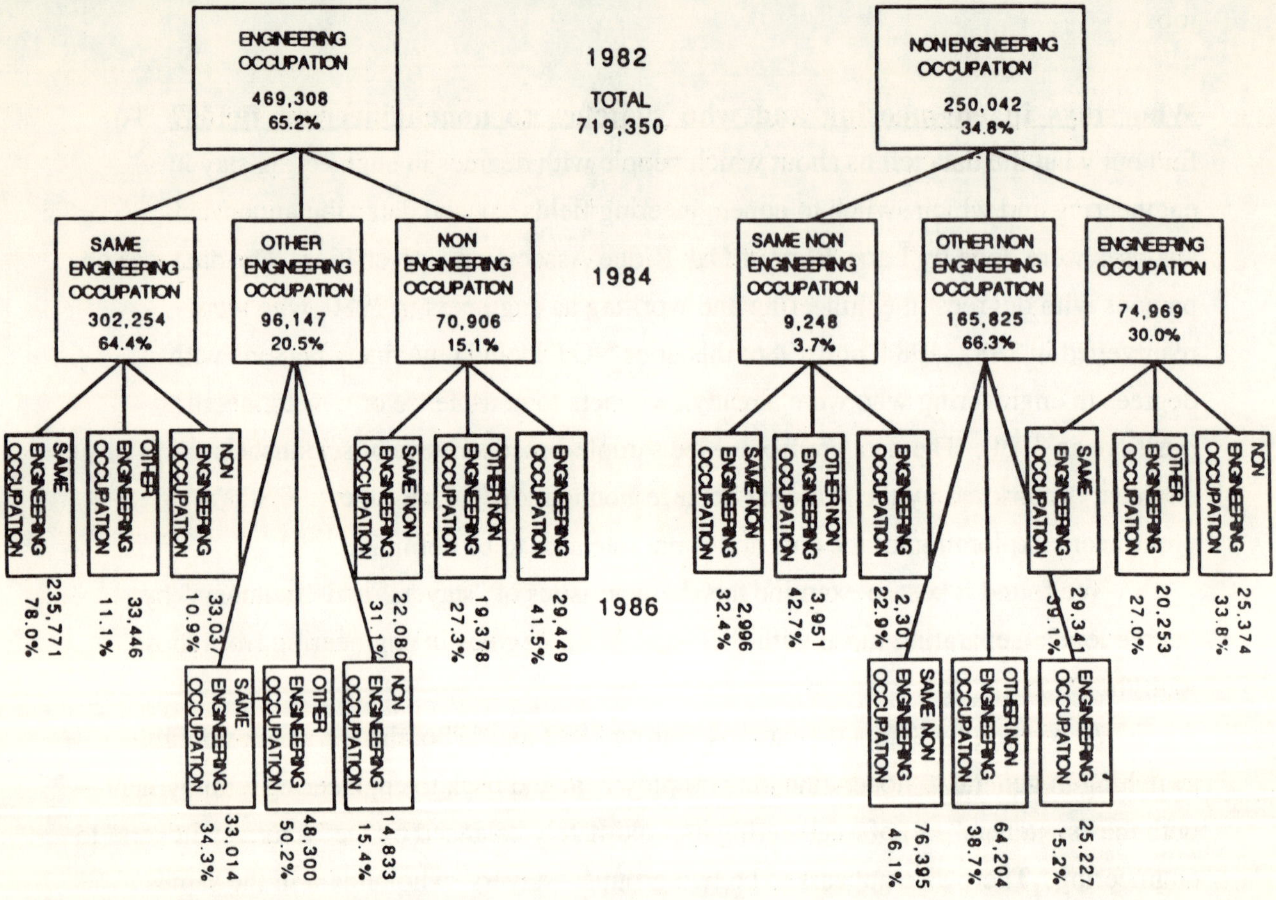

Employed and reporting in 1982, 1984, and 1986.
Missing occupation responses are included in non-engineering.
Sample size = 12,962 or 1.8%

SOURCE: Labor and Policy Studies Program, Science/Engineering Education Division, *Engineering Mobility and Salary Information*, Oak Ridge, Tenn.: ORAU, 1989, based on 1986 postcensal survey, strata 1-10.

Figure 1. Individuals holding B.S., or M.S., engineering degrees in 1980, by employment specialty, 1982,1984,1986.

Ph.D. "switchers" out of engineering are almost offset by the "return switchers" back to engineering (Figure 2). Switching does not appear to be related to years since the Ph.D. in engineering was earned, nor to age. Among those with the Ph.D. in engineering, "switchers" are somewhat more likely to be employed in business and industry and government and much less likely to be employed in education institutions. Ph.D. "switchers" compared to "stayers" by primary work activity have lower percentages in teaching and R&D, about the same percentage in R&D management, and higher percentages in other management and operations/other.

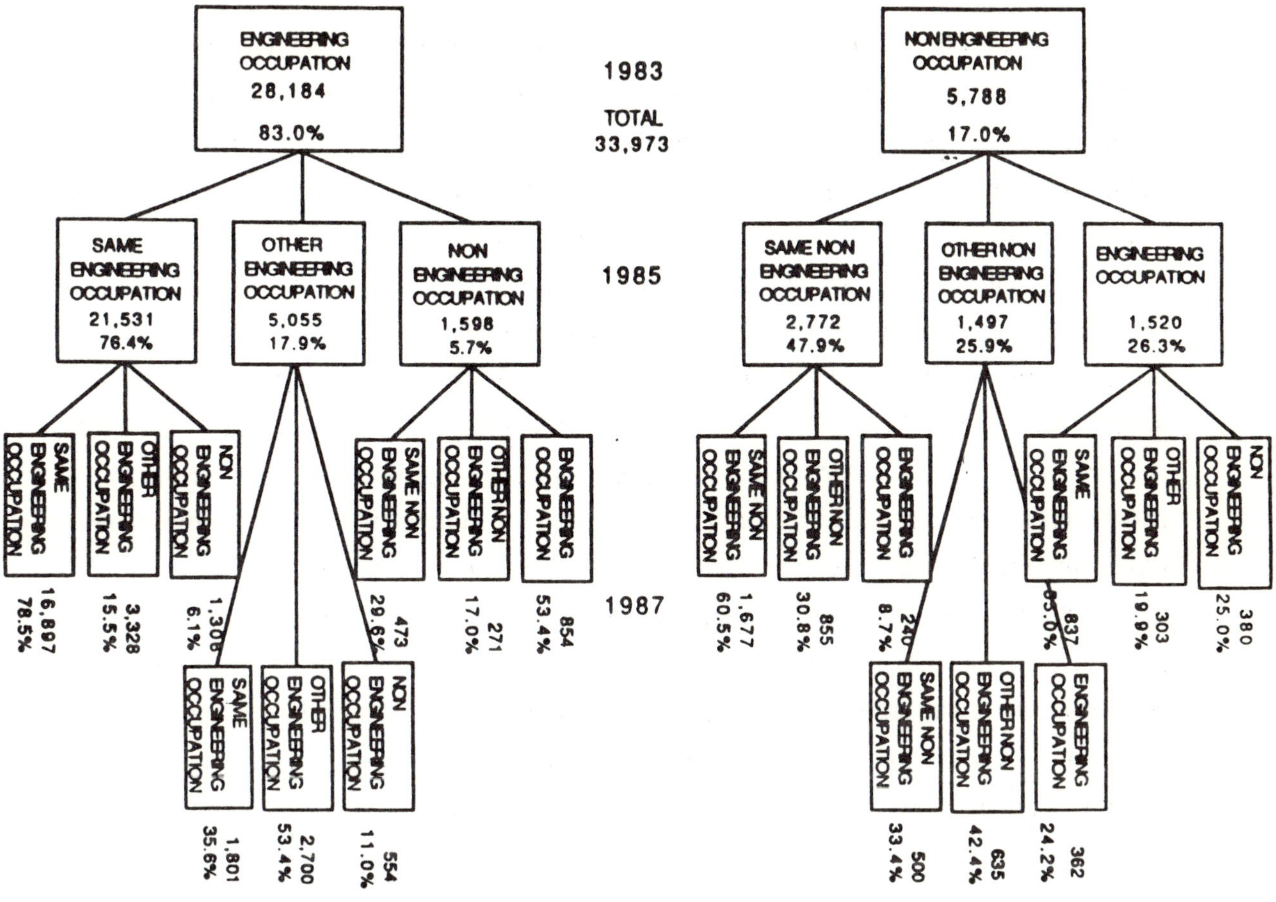

Employed and reporting in 1983, 1985, and 1987.
Missing occupation responses are included in non-engineering.
Sample size = 1,070 or 3.1%.

SOURCE: Labor and Policy Studies Program, Science/Engineering Education Division, *Engineering Mobility and Salary Information*, Oak Ridge, Tenn.: ORAU, 1989 based on the 1987 longitudinal doctorate survey.

Figure 2. Individuals holding Ph.D.s in engineering in 1983, by employment specialty, 1983, 1985, 1987.

Are there two career tracks—technical and managerial? Blair's analysis indicates that there appears to be two tracks at the B.S., and M.S., levels:

- The average salaries for those in nonengineering employment are 13 percent higher than the average salaries for those in engineering employment.
- The highest paying primary work activities are R&D management and other management where salaries for those in nonengineering employment are substantially higher than for those in engineering employment.

For those with B.S., or M.S., degrees, having a graduate business degree adds greatly to salary; working in science and engineering (S&E) occupations as compared to non-S&E occupations decreases average salary levels. In sum, Blair's analysis indicates that for holders of B.S., and M.S., degrees in engineering, the technical track is inferior in terms of earnings; in other words, there really are not two tracks.

Similarly, there does NOT appear to be a dual career track for those with the Ph.D. in engineering. R&D management and other management categories show substantially higher salaries for those with Ph.D.s in engineering for both engineering and nonengineering employment specialties. In fact, average salaries in the management areas are somewhat higher for those indicating an engineering employment field. The only exception is in operations/other primary work activity, where nonengineering employment has a substantially higher average salary than engineering employment.

Studies have shown that imbalances between supply and demand do not lead to crises; we do not have unfilled jobs, but jobs filled by people from other areas. Can American society afford to wait on unassisted market mechanisms or do we want to do something to try to dissipate changes and move things in a direction that policy wants them to go?[7] This is the starting point for the workshop discussions.

References

Atkinson, Pamela H. 1989. *The Relevance of Career-Long Education to Creating and Maintaining an Adaptable Work Force*. Paper presented at the Workshop on

[7]In this context, policy not only refers to that created by the federal government, but also to policy devised by other institutions such as economic and education institutions.

National Needs and Technological Change: Fostering Flexibility in the Engineering Work Force, National Academy of Sciences, Washington, D.C., September 29.

Cyert, Richard M., and D. C. Mowery (eds.). 1987. *Technology and Employment, Innovation, and Growth in the U.S. Economy*. Washington, D.C.: National Academy Press.

Eck, Alan. 1989. *Adaptability of the Engineering Work Force: Information Available from the Bureau of Labor Statistics*. Paper presented at the Workshop on National Needs and Technological Change: Fostering Flexibility in the Engineering Work Force, National Academy of Sciences, Washington, D.C., September 29.

Gorn, Saul. 1989. *Adapting to Computer Science*. Paper presented at the Workshop on National Needs and Technological Change: Fostering Flexibility in the Engineering Work Force, National Academy of Sciences, Washington, D.C., September 29.

Watson, J. S. 1989. *Adaptability in Chemical Engineering*. Paper presented at the Workshop on National Needs and Technological Change: Fostering Flexibility in the Engineering Work Force, National Academy of Sciences, Washington, D.C., September 29.

Woodall, David M. 1989. *Nuclear Engineering Case Study*. Paper presented at the Workshop on National Needs and Technological Change: Fostering Flexibility in the Engineering Work Force, National Academy of Sciences, Washington, D.C., September 29.

APPENDIX A
EDUCATION FIELDS OF EARLY PH.D. NUCLEAR ENGINEERS

Larry Blair
Oak Ridge Associated Universities

Evidence from the 1975 Survey of Ph.D. Scientists and Engineers

- The majority of early Ph.D. nuclear engineers earned their degrees in physical science disciplines (about 7 out of 10), with physics being by far the most common area of study. This is apparent in the 1975 longitudinal Ph.D. survey data,where almost 70 percent of the over 55-year-old age group report physical science as their major field of study. Approximately 50 percent of the 45- to 55-year-old age group indicate physical science fields of study, but less than 25 percent of the under-45 age group report physical science fields of study.
- Engineering other than nuclear was the field of study for 30-40 percent of the 45-and-older age groups.
- Dramatically, none of the over-55 age group and only 5 percent of the 45-55 age groups had a nuclear engineering major as a field of study. However, among the under-45 age group, over 50 percent had a nuclear engineering major in their Ph.D. studies.

Evidence from the 1987 Survey of Ph.D. Scientists and Engineers

- The trend toward a nuclear engineering major in Ph.D. studies and away from a physical science field also is clearly seen in comparing Ph.D. nuclear engineers in the 1987 survey to those in the 1985 survey. By 1987 the percent of the under 45-year-old age group with a nuclear engineering major had decreased from approximately 50 percent to almost 70 percent. The percent of the 45-55 age groups with a nuclear engineering major increased over four times, to approximately 40 percent, and even a small percent of the over-55 age group has a nuclear engineering major.
- Conversely, the percent of nuclear engineers with physical science majors decreased by 1987, dramatically for all age groups 55 and younger. The under-45 age group with a physical science major decreased from more than 20 percent in 1975 to less than 10 percent in 1987, the 45-55 age groups from approximately 50 percent to less than 20 percent, and even the over-55 age group from almost 70 percent to 60 percent.
- The percent of Ph.D.s employed as nuclear engineers but holding degrees in other engineering majors was about the same in 1987 as in 1975 for each of the age groups.

Table 1. Ph.D.s Employed as Nuclear Engineers in 1975 by Degree Field and Age, 1987 Longitudinal Doctorate Survey.

	Age Group (Percent Distribution)			
	Under 45	45-49	50-55	Over 55
Ph.D. Degree Field				
Physical Sciences	23.7	47.6	52.4	69.6
Engineering				
Nuclear	52.8	6.6	4.2	0.0
Chemical	6.8	16.9	21.5	18.4
Other	14.6	27.7	17.8	12.0
All Other	2.1	1.2	4.2	0.0
Total	100.0	100.0	100.0	100.0

	Age Group (Weighted Numbers)			
	Under 45	45-49	50-55	Over 55
Total	1124	166	191	158
Math/Stat	20	2	0	0
Physics	197	67	88	87
Chemistry	69	12	12	23
Engineering				
Aero/Astro	23	0	0	0
Bioeng/Biomed	15	0	0	0
Chemical	76	28	41	29
Elec/Electron	10	0	2	0
Nuclear	594	11	8	0
Eng Mech	20	11	11	0
Eng Physics	53	0	0	0
Mechanical	33	26	21	0
Metallurgical	10	0	0	19
Fuel Tech	0	9	0	0
All Other Fields	4	0	8	0

Sample size = 178 or 10.9%

Table 2. Ph.D.s Employed as Nuclear Engineers in 1987 by Degree Field and Age, 1987 Longitudinal Doctorate Survey.

	Age Group (Percent Distribution)			
	Under 45	45-49	50-55	Over 55
Ph.D. Degree Field				
Physical Sciences	7.2	13.9	17.8	60.9
Engineering				
Nuclear	69.2	43.6	39.2	2.3
Chemical	0.1	6.3	21.4	13.5
Other	23.5	35.1	17.2	23.4
All Other	0.0	1.1	4.5	0.0
Total	100.0	100.0	100.0	100.0

	Age Group (Weighted Numbers)			
	Under 45	45-49	50-55	Over 55
Total	891	447	337	394
Math/Stat	0	0	15	0
Physics	59	62	60	152
Chemistry	5	0	0	88
Engineering				
Aero/Astro	0	53	0	0
Bioeng/Biomed	30	0	0	0
Chemical	1	28	72	53
Civil	50	0	0	0
Elec/Electron	27	0	28	66
Nuclear	617	195	132	9
Eng Mech	2	0	30	0
Eng Physics	0	104	0	0
Mechanical	92	0	0	18
Gen/Other	8	0	0	8
All Other Fields	0	5	0	0

Sample size = 88 or 4.3%

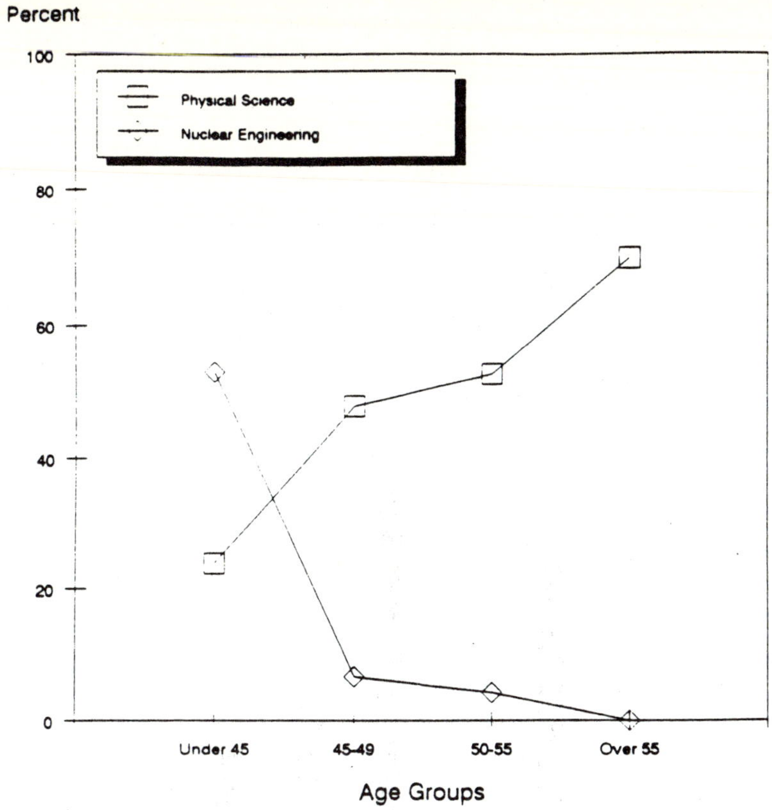

Figure 1. Ph.D.s employed as nuclear engineers in 1975, physical science versus nuclear engineering, fields of study in Ph.D. by age group.

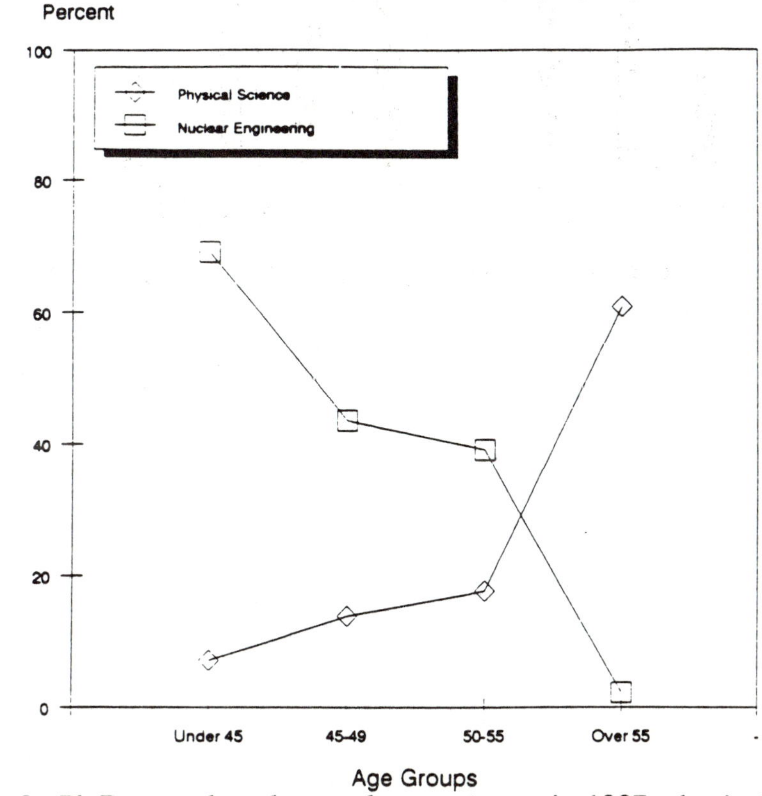

Figure 2. Ph.D.s employed as nuclear engineers in 1987, physical science versus nuclear engineering, fields of study in Ph.D. by age group.

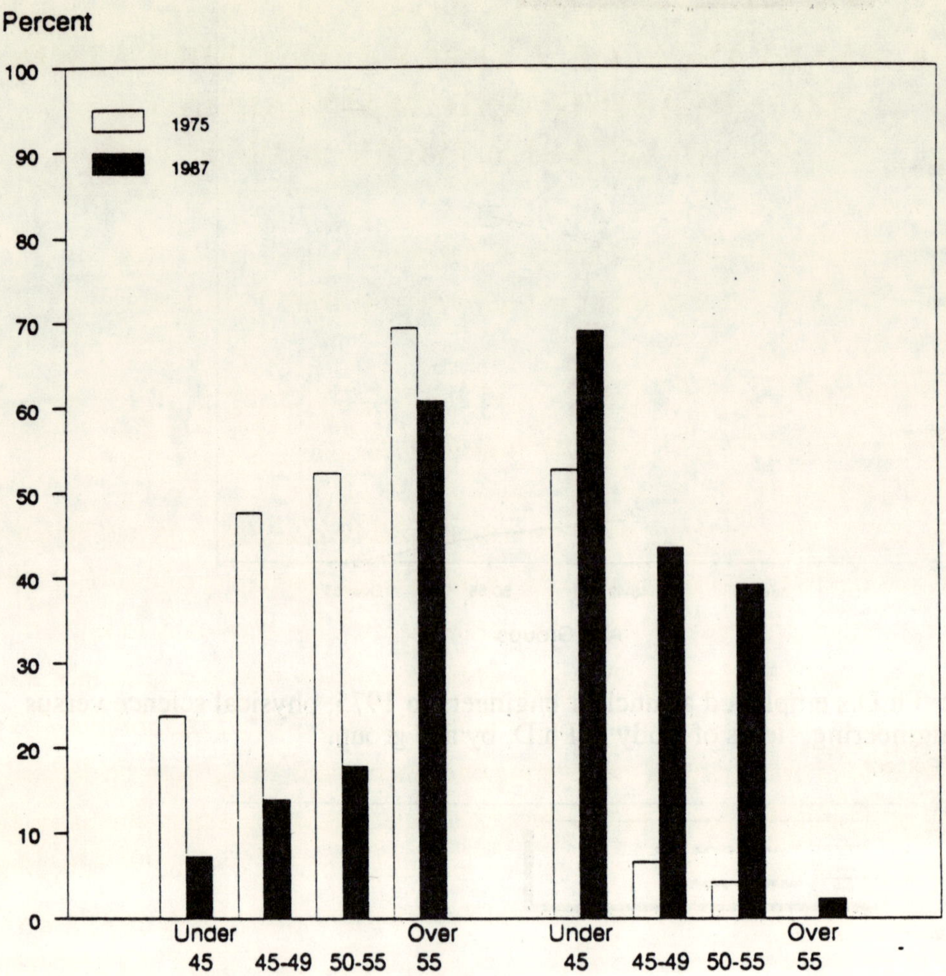

Figure 3. Physical science and nuclear engineering majors, 1975 vs. 1987 for Ph.D.s employed as nuclear engineers by age group.

THE RELEVANCE OF CAREER-LONG EDUCATION TO CREATING AND MAINTAINING AN ADAPTABLE WORK FORCE

Pamela H. Atkinson
University of California, Berkeley

Introduction

Those who are concerned about the nation's ability to develop and maintain a competent and flexible engineering work force believe that continuing education is a major contributor to engineering adaptability. No actual research has been done to support this thesis, however; and anecdotal information, while impressive, provides few clues as to how, when, and to what degree new or continuous learning affects adaptability.

Career-long or continuing engineering education and training[1] may be defined as knowledge or training in the use of new processes, systems, devices, and chemicals that will improve an engineer's ability to do his/her job. Additionally, continuing education can relate to competency as well as to professional development. Career-long competency is the ability of an engineer to remain competitive, productive, and creative as a peer-respected member of the engineering profession through the acquisition and application of the new knowledge, skills, and experience that a changing profession demands. Continuous professional development is activity that upgrades, expands, and renews the competence of an engineer throughout a professional career, enabling him/her to adapt to changing times while remaining an active contributor.

To be adaptable means to be willing to make changes—to adapt oneself to new situations and the changes in one's environment. For the purpose of discussion,

---

[1]The terms "education" and "training" are used somewhat interchangeably by industry without the distinction of a value difference. Academe, on the other hand, perceives "training" as more functional than knowledge-based and less intellectual, therefore of lesser value and tends to not use it as a descriptive when discussing the subject. Pertinent to any discussion of this issue is the fact that engineers employed by the armed services are governed by a bureaucratic system that only allows "training" to be reimbursable, not "education." For that reason, any discussion of career-long learning that affects the national cadre of engineers needs to include the word "training" as well as "education."

"environment" should be understood to mean many things: one's personal environment in terms of work station and job responsibility; a broader interpretation, such as all the divisions of a company and how they interface; even a complex environment, such as the various spheres of activity and influence of a multinational corporation. The ways in which a company adapts to change are based on decisions made within the company by its officers and its management personnel. Often the need to change inspires attitude shifts that are philosophic in nature, based on practical concerns, and influenced by reasoned judgment. Policy changes made at the highest level affect everyone in the company; therefore, the thought processes of decisionmakers should not be exempt from study, analysis, and interpretation.

Systematic research needs to be conducted to identify the link between continuing education and adaptability. And the subjects of that research should include not only the engineering work force, traditionally line and bench engineers and technicians, but also project and group leaders, upper-level management, and CEOs.

This paper provides a quick overview of the issues central to the subject of continuing education in engineering. These issues are numerous, the problems they highlight are real, and there are no quick fixes that will provide long-term, across-the-board solutions; most of these issues have been discussed in workshops such as this one since the 1950s. Additionally, this paper presents a number of ideas and questions that may be useful to explore, if guidelines for identifying the link between continuing education and adaptability are to be designed, approved, and implemented.

Continuing Engineering Education is a Human Resource Issue That Prompts Continual Studies

A company's ability to adapt to the changing demands of the marketplace is related to its ability to manage money, projects, and people and to plan for and respond to the exigencies of business and the dynamics of technological opportunity in pursuit of long-term goals. Of these tasks, the management of people may be the task most critical to a business's success and the one hardest to assess, evaluate, and plan for in a systematic way. In 1987 and 1988, the intensely volatile marketplace and the pressure on companies to turn out quality products ahead of their competitors—and at a lower price to the consumer—made the technical training and updating of engineering professionals the "hot topic" on the national engineering agenda, since the company that employs the most up-to-date technical work force is the company that is able to use technology to improve its market advantage and its competitive position. A report by the National Academy of Engineering (1988) presented a series of recommendations to companies, universities, government agencies, and professional societies on ways to combat engineering obsolescence in the workplace and in academe. Many of these recommendations were similar to recommendations advanced in previous reports; for example, the 5-year study by the American Society for Engineering Education (ASEE, 1968); the report *Lifelong Cooperative Education* of the Department of Electrical Engineering and Computer Science, Massachusetts Institute of Technology (MIT, 1982); and the NRC comprehensive-volume study on engineering education and practice (1985). A similar report issued by ASEE in late 1987 addressed the total engineering education enterprise and the need to develop innovative plans throughout its sectors to strengthen all aspects of engineering education, including educational instruction for faculty to bring them up to date with new technologies.[2]

The Engineering Work Force: A Numbers Problem? Or a Utilization Problem?

Because of the changing demographic face on the U.S. work force, companies can no longer expect to continue the custom of hiring significant number of new engineers every year. To a great degree, they will have to either lure talented professionals from their competitors or retrain and upgrade the professionals they have. Of these two strategies, retraining and upgrading have the greatest potential for the long term and positive ramifications for achieving national economic goals (which cannot be said of the first strategy).[3]

Briefly, the demographic issues of concern to the engineering community are these:

- The country is experiencing a "baby bust," a decline in the college-aged population;

[2]For a chronological review of engineering education reports since 1918, see Appendix A, compiled by Daryl Chubin for *Higher Education for Science and Engineering: A Background Paper*, Washington, D.C.: Congress of the United States, Office of Technology Assessment, March 1989.

[3]Three-quarters of the current work force will be employed in the year 2,000; within engineering and engineering-related activities alone, that number represents approximately 2,168,000 people and a sizable economic investment in human resources.

- The number of undergraduates enrolled in engineering degree programs has been dropping since 1986, and Americans' apparent lack of interest in graduate engineering programs continues;
- The large number of foreign students in graduate programs is rising, and a great number of them will be hired as engineering school faculty; and
- Foreign faculty's views on woman's appropriate place in society may create problems for women graduate students and adversely affect the ability of graduate engineering programs to retain women students at a time when their numbers are needed (Vetter, 1988).

Although these demographic indicators cause some people to believe that a shortfall of engineers is a real possibility in the near future, history shows us that there is a considerable amount of flexibility in the engineering work force—a self-correcting mechanism, if you will—and shortages may apply only to specific sectors before the self-correcting mechanism springs into action. It is exactly this mechanism, and how it fine-tunes itself, that we need to learn more about. The Office of Technology Assessment (OTA, 1985) noted,

> There is no national market for scientists and engineers as a group. Rather, there are specific markets for graduates trained in particular disciplines, and these markets and disciplines can experience very different conditions at the same time. For example, the National Science Foundation (NSF) projected in 1982 that the demand for electrical and aeronautical engineers and for computer specialists could exceed the supply of graduates in those fields by as much as 30 percent over the next 5 years, while at the same time there would be significantly fewer openings for biologists, chemists, geologists, physicists, mathematicians, chemical, civil and mechanical engineers than there would be trained degree-holders.[4] Thus, it is individual disciplines, especially those linked with high growth or defense-oriented industries, that could experience personnel problems in the future—not "science and engineering" as a whole. . . [Additionally], college students appear to be

[4]That has not been the case in industry; essentially supply and demand have worked in concert in the last few years.

> highly responsive to market signals and appear to shift their career choices dramatically toward fields that promise greater occupational rewards.

OTA concludes that

> career choices and market forces have a greater impact on the supply of scientists and engineers than do demographic trends. . . [except in academe], however, where known demographic trends exert considerable influence.

An Exploration of the Issues

The following is a synthesis of many of the questions and issues raised by the engineering community over the last few years in respect to career-long education for engineering professionals.

Continuing Engineering Education in the Workplace

What career-long education and training would optimize the knowledge, experience, and proficiency of the engineering professional as well as insure more technical flexibility? Ideally, it would be continuous exposure to and training in the use and management of the latest materials, chemicals, systems, devices, and processes as they are developed and improved.

What does continuing engineering education look like? It comes in a variety of shapes and sizes and a number of infinite designs (for large or small groups, even for individual self-study). It is delivered in various ways, from the traditional classroom that instructs 30 to the national satellite broadcast that informs thousands. Some forms of career-long education are

- in-house technical training;
- interactive video-based training;
- self-paced computer-based training;
- in-house "expert" seminars and intensives;
- job-related experience ("learning by doing");

- reading technical reports and journal articles;
- university extension short-courses and seminars;
- national and international conferences and meetings;
- external professional programs leading to a certificate;
- external academic degree programs leading to the Ph.D.; and
- external academic degree programs leading to the master's degree.

Who Pays for Continuing Education and the Issue of Part-Time versus Full-Time Education

Corporate industry continues its long-established tradition of not only providing education and training for its employees, but paying for it as well.[5] Some predict that the training and education of some 40-50 million workers will become the largest industry in the United States in the 1990s: the number of changes in the workplace, such as remedial training of employees in the use of new technologies and the shift in management cultures to quality methodologies, will lead to "explosive increases" in the demand for adult education, doubling or tripling the cost of employee training in the next five years (Chappel, 1989). Industry now supports education in a number of ways, such as:

- tuition reimbursement for credit courses and degree programs (some companies also pay for audit courses);
- in-house training programs by outside experts as well as off-site workshops, conferences, and seminars;
- tutorials and intensives offered by in-house experts;
- mentor and tutoring programs;
- travel expenses and per diem for employees to conferences;
- career-counseling; and
- individual guides for self-study.

---

[5]Although estimates of the cost of employee-funded education programs for technical staff vary, there is agreement that the cost is in excess of $30,000,000,000 a year and accelerating. Very few companies know EXACTLY how much their education programs actually cost, because the cost of salary is rarely factored in. In addition, many long-term education programs are perceived philosophically as "not possible" by some companies, who budget for them not year-to-year, but quarter-to-quarter!

While education and training is considered by most companies to be as crucial to success as research and development, and many large companies allocate the same percent of operating money to it, education funds are historically the first cut and the last reinstated when times are tough; long-term objectives are hard to remember when stockholders have short-term expectations. And for the small company, the governing rule is that the individual is on his/her own; that is to say, it is the personal responsibility of every engineer to stay current in his/her field. Sometimes education is reimbursable, but more often it is not, because there is no money set aside to pay for it.

In almost all discussions of continuing education programs financed by companies, the focus of discussion is the education of large numbers of engineers through in-house programs or off-site degree programs that accept working engineers part-time. In addition to these programs, a number of companies—such as AT&T Bell Labs, General Electric, Hewlett-Packard, and IBM—send a limited number of people to school full time, at both the master's level and the doctoral level. During this time, the company normally pays 50-100 percent of their salary (in addition to tuition, books, and a cost-of-living stipend) and maintains their benefits as if they were actually employed. These programs are small, select, very competitive, and, by and large, outside the means of small and mid-sized companies.

A number of prestigious institutions—such as Stanford, the University of Illinois at Urbana-Champaign, the University of Southern California, Purdue University, the University of Washington, and Carnegie-Mellon University—allow engineers employed in industry to enroll in regular on-campus master's programs on a part-time basis; some of these schools establish off-site degree programs as well. Similar programs are also offered by other 4-year institutions, public as well as private, but many universities have chosen to stay on the sidelines.

Graduate School, One Aspect of Continuing Education

Industry recruiters who compete for top engineering graduates emphasize tuition-reimbursed degree programs as a benefit of employment. The high cost of a graduate education and the lack of enough funding to support large numbers of graduate students in research associate positions makes these company-sponsored degree programs very attractive to B.S. graduates. As time becomes more precious, highways become more congested, the need for technical expertise becomes more acute, and graduate programs become more expensive for the individual to fund, more and more B.S. graduates will opt to defer graduate education until they are employed full time and let their employer pick up the tab.[6] The number of students enrolled part-time in master's degree programs in the fall of 1988 totaled 29,975, or 37 percent of the total enrollment of 79,997 (ASEE, 1989). As the need for technical currency escalates, and if American students continue the trend of rejecting graduate education in engineering after attaining the B.S. degree, it is likely that this part-time segment will increase.[7]

Educational Entrepreneurs Abound

Companies demand high-quality instruction geared to adult professionals, delivered in a timely fashion in a cost-efficient, cost-effective manner. A number of educational entrepreneurs are responding to this need, and if they continue to provide the high-quality service they are becoming known for, their programs will continue to be much in demand.

Those providers that utilize television satellite and microwave technologies to deliver on-site programs to industry are meeting a very real need, and industry has been generous in its support of their efforts. Although Rensselaer Polytechnic Institute, Purdue University, the University of Washington, Stanford University, and Chico State University are just a few schools out of dozens that use television to transmit their engineering degree programs to local industry, the provider that offers the most promise for the future is the National Technological University, a consortium of 29 universities offering a number of

---

[6]HR Section 127 is the most recent proposal to challenge the value and effectiveness of company-funded engineering education programs. This legislation, approved until the end of 1991, imposes a tax to the consumer—the individual engineer—on company-paid-for education programs. As of August 1989, this bill allows employers to pay up to $5,250 per person in education benefits without employees having to pay taxes on the funds. The flaw in this approval is that it only applies to undergraduate programs. There is currently a major effort under way to restructure the wording in this bill so that graduate education is part of the provision.

[7]The disaggregation numbers of degree-recipients each year does not separate full-time from part-time grantees. Although these two cohorts are tracked separately as students, once they earn the degree the distinction that separates them vanishes. Because current statistical research looks at the number of B.S. students that graduate every year and tracks how many of them go to graduate school, those that enroll at a later date through their company's education program fall through the cracks. They appear in the data 2-5 years later as part-time students, but that information is itself misleading, because it counts them as members of the current graduating cohort. Data specify the number of part-time students who attain graduate degrees would give valuable information about who drops out of graduate programs—the full-time, younger student or the older, professional engineer.

engineering, computer science and management degree programs to over 242 industry sites via a television satellite network.

Yet Another Study on Education and Training in the Work Place

The American Society of Training and Development (ASTD) is conducting its own 2-year study on personnel issues and education and training in the workplace. The report of that study is scheduled for distribution early in 1990 through Jossey-Bass publishers. Some within ASTD believe that it is the B.S. engineer who is becoming obsolete, that technicians will be assigned jobs once held by B.S. engineers, and that in the future those who want challenging jobs will have to hold master's degrees. (This is not a new idea: the master's degree was predicted to soon become the basic engineering degree in the 1968 ASEE "engineering goals" report, but this has not happened; the B.S. in engineering is still the recognized entry-level professional degree.) The question of B.S. degree versus M.S. degree and entry into the work force as an engineering professional raises the issue of the value of the Master of Engineering degree and the Doctor of Engineering degree, with their emphasis on engineering practice, and whether these degrees are more appropriately suited to engineers in industry than are the Master and Doctor of Science degrees, which may be more research-oriented.[8]

Looking to the Future

Even as rapid improvements are being made in the design and application of new technologies, the depth of engineering disciplines in the universities is increasing, broadening the knowledge base that engineers of the future will have to have to practice their profession; this in turn will add to the number of knowledge elements they will need to keep current. The IEEE Society, concerned about a way to bridge the knowledge gap of

---

[8]One side argues that a practice-oriented degree does not produce an engineer with vision, and therefore will never replace in value the M.S. and Ph.D. degrees. Another side maintains that a practice-oriented degree incorporates the elements of cost, reliability and timeliness, as well as theory, into engineering design in a comprehensive way and believes that those elements are not only important but essential to good practice.

those currently in the workplace and prepare them for the adjustments they will have to make as changing engineering practice requires new professional skills, is working with groups of experts to create self-administered tests/questionnaires that will show what field-specific knowledge elements one would need to move into certain new areas—for example, moving out of magnetics into fiber optics, or out of heat-transfer into thermodynamics.[9] Once the value of such a test is determined, other tests can be designed for other disciplines and long-term plans can be made to revise the questionnaires as needed to keep them current. What is still to be discovered is how many engineers will make use of these self-administered tests and obtain the knowledge elements that a new field would require them to have; and how many will take the tests, find themselves severely limited by their lack of new knowledge, and choose instead to leave engineering and find other work rather than opt for additional education. The question of incentives and support systems to encourage new learning is crucial and has to be explored in depth.

The Current Canon

While current education and training needs of companies appear to be met by the variety of educational providers in the marketplace, many responsible leaders within the engineering community are concerned about the lack of leadership in this field. Some academics see current training programs nationwide as more reactive than proactive, concerned more with fixing immediate problems than preventing future ones. They decry the lack of structure, quality control, and program planning, yet paradoxically refuse to be directly involved in improving the status quo of that which they disdain. The industrial point of view is succinct and to-the-point: companies want high quality, need-specific, well-produced programs designed for adult learners. They want programs that can be delivered in a timely fashion and at a cost they can afford: they are concerned to get the best bang for their buck and they shop around. A number of individuals within both groups believe that a concerted effort of all the players is needed to assess and plan for the future education and training needs of engineers in industry; they believe that action, not

[9]This test (see Appendix B), still in a pilot stage, was inspired by the "Self-assessment Procedures for the Computer Professional," a self-administered test published by the *Journal for the Association of Computing Machinery*.

further study, is what is needed. They also believe that this national effort is imperative if the United States is to hold its own against its competitors.

Some Questions for Discussion

- Many major research institutions have equally prestigious schools of engineering. These engineering schools, while concerned about the need for continuing education of engineers in industry, choose not to be involved and say that it is industry's problem, not theirs. Are there any new persuasive arguments that might encourage these schools to contribute their talents to the national effort?
- Could better counseling of undergraduate engineering students help inculcate a sense of personal responsibility for continuous education and training after graduation?
- Could systematic on-the-job counseling, including annual performance reviews, encourage working professionals to participate in education and training on a regular basis?
- Mid-line and project managers are often cited as frequent obstacles to the professional development of technical staff. Some companies find that staff education and training programs succeed when managers enroll as participates and, additionally, when managers are rewarded by how well they develop their own people. Are hard data available to support this thesis?
- Are there new and improved ways that might link educators and practitioners and improve the professionalism of both?
- Are there opportunities that might naturally bring the engineering professional back to the campus for short periods of time and, conversely, the academic into the industrial environment? The model of teacher as trainer as well as the model of engineer as teacher have both been successful; why is it that these models are so underutilized?
- Does co-op experience make an engineering professional more adaptable? If it does, are their ways to expand the number of co-op opportunities available to students?
- Is any one year within a 4- or 5-year program better than any other year for a student to participate in co-op?

- Is there a need for a carefully-designed and manufacturing-oriented graduate degree program in the field of management of engineering resources? Or should this be exclusively each individual company's responsibility? (This would include more than just the management of technology, since for many practicing engineers, career-long education at some point needs to prepare them as managers of product development, product cycles, and people.)
- Are there do-able modifications/additions/changes to current undergraduate engineering programs and tracks that might improve a young engineer's ability to adapt to the changing professional demands of the workplace plus increase his/her worth in a fast-changing, technologically sophisticated marketplace?
- Might it be of mutual benefit for professional schools to establish a strong tie to their alumni by offering summer intensives, special courses, or even a newsletter that helps guide them as they prepare to move into new career paths? How might alumni be developed as a resource separate from that of fund-raising?

Imaginary Profile: Tomorrow's Engineering Schools Prepare Engineers for Adaptability

Tomorrow's engineering graduate will be even more technically competent than is now and will be skilled as a communicator of ideas, both in written and oral form. Heavy emphasis on laboratory courses in conjunction with a comprehensive background in calculus and physics have broadened the engineer's ability to put theoretical constructs into practice. Engineering schools emphasize the importance of quality control, design, adherence to project schedules, and concern for cost and reliability as basic to good engineering practice. Numerous oral presentations and written reports are required for all courses, and the accepted project approach is to work in teams.

Interest in and support of engineering co-op programs has soared, and companies fund not only co-op appointments for students, but also specialized training and financial support for co-op directors and their staff, linking co-op to career-counseling and professional development at both ends of the spectrum. All schools require that their students have at least one co-op appointment.

Fellowships to support undergraduate research projects are numerous. These undergraduate fellowships awaken student interest in Ph.D. programs and an academic career, and fully paid graduate scholarships (renewable based on merit and promise) attract many who might not otherwise be encouraged to enroll in graduate school.

Can Engineers Learn to Adapt to Meet Changing Engineering Employment Needs?

Today's engineers need to assume responsibility for their own professional development, if they are to be adequately prepared for the complex challenges of tomorrow's workplace. Technical proficiency kept sharp through education and training is a key to being responsive to and ready for the changes in engineering employment. The following professional development strategies directed to beginning engineers seem to encourage adaptability:

- As soon as you are able, find a mentor. When/if you change companies, implement this strategy first. Having a mentor is a crucial component of getting on and staying on a successful career path; it is probably the most important action as you begin professional practice.
- Once hired, locate the education and training department of your company as soon as possible and ask for literature describing the company's educational policies and programs.
- Identify the program, courses, and seminars that will benefit you the most AND increase your value to the company.
- Consistently throughout your employment, take advantage of as many education and training opportunities as possible.
- Keep informed about your company and know the role your particular department/division plays as one part of a larger whole. Learn as much as you can about the company history, philosophy, and goals.
- Volunteer to be a mentor for a new employee, to sit on the education and planning committee, and to serve on community projects. Be visible.
- Read as much as you can about the global economy and world affairs to gain a perspective of the company's financial future. Keep abreast of the latest technologies in your own field. Learn to relate your educational needs to the company's long-term goals.

- Join a professional society and play an active part in it. Keep up with the literature.[10]

References

American Society for Engineering Education (ASEE). 1968. *Goals of Engineering Education*. Washington, D.C.: ASEE.

-----. 1987. *A National Action Agenda for Engineering Education*. Washington, D.C.: ASEE.

-----. 1989. *Engineering Education* 79(March):72.

Chappel, S. 1989. Stepped up training of U.S. workers: a costly necessity. *Engineering Times* 11(October):1.

Department of Electrical Engineering and Computer Science. 1982. *Lifelong Cooperative Education*. Cambridge, Mass.: Massachusetts Institute of Technology.

National Academy of Engineering (NAE). 1988. *Focus on the Future: A National Action Plan for Career-Long Education for Engineers*. Washington, D.C.: NAE.

National Research Council, Committee on the Education and Utilization of Engineers. 1985. *Engineering Education and Practice in the United States: Foundations of Our Techno-Economic Future*. Washington, D.C.: National Academy Press.

Office of Technology Assessment (OTA). 1985. *Demographic Trends and the Scientific and Engineering Work Force: A Technical Memorandum*. Washington, D.C.: U.S. Congress.

Vetter, Betty M. 1988. *Demographics of the Engineering Student Pipeline*. Paper presented to the ASEE Engineering Deans' Council, Washington, D.C., January 7.

[10]The proliferation of reports, articles, and professional papers in some disciplines may be of such volume that the last recommendation is impossible. For example, the number of professional papers in chemical engineering journals today nearly doubles the number of papers published in journals in the 1950s. Some chemical engineers estimate that discoveries in their field are moving so rapidly that it won't be long before the knowledge base of a chemical engineer begins to erode six months after graduation from college!

APPENDIX A*

1918 Publication of the Mann report, *A Study of Engineering Education*, sponsored by the Society for the Promotion of Engineering Education (SPEE) and funded by the Carnegie Foundation. It urged return to fundamentals and unify fragmenting curricula; merge theory and practice in coursework; introduce "real work," including "values and costs," into teaching engineering problem solving; retain shop experience, laboratory, industrial training, cooperative and summer work in curriculum; English mastery; link technology to its human and social setting; closer university-industry linkage, especially in research, to improve productivity and thereby national well-being; develop discipline for work and "lifelong" study; and select faculty based on teaching ability and work experience, not just research excellence.

1930 Publication of volume 1 of the Wickenden Report, *Report of the Investigation of Engineering Education 1923-1929*.

1934 Publication of volume 2 of the Wickenden Report. It urged a halt to fragmentation of curricula; graduate engineering education and continuing education for 5 years after graduation; forms of technical education other than engineering colleges; functional rather than professional engineering education; design project, including writing, for second- and third-year students; third-year project teaching, fourth-year honors option; stronger high school preparation; lifetime learning in cooperation with industry; professional certification by engineering societies independent of State licensing; higher faculty standards; teach engineering method; teach society and values so engineers can understand social impact of engineering.

1939 H. P. Hammond, report for SPEE, *Aims and Scope of the Engineering Curriculum*, recommended diversification of curricula; parallel technical and humanities/social sciences "stems"; reconsideration of 4-year curriculum and move to 5- or even 6-year program.

1944 H. P. Hammond, report for SPEE Committee on Engineering Education after the War, reaffirmed 1939 report; promoted expanding technician programs to fill industrial needs then being met, non-optimally, by engineers; and teaching the "art" of engineering as distinct from scientific method.

1955 L. E. Grinter, *Report on the Evaluation of Engineering Education* for American Society for Engineering Education (ASEE). The final report included comments by 122 engineering colleges. It recommended: five "stems"—humanities and social sciences, mathematics and basic science, generic engineering science, engineering specialty subjects, and electives; a two-track undergraduate curriculum, one to immediate employment, the other to graduate study; twin goals for engineering education—technical (analysis and "creative design"; construction, production, operation) and social (ethics, general education, leadership in technological action); improved high school preparation

---

* SOURCE: Steven L. Goldman, "The History of Engineering Education: Perennial Issues in the Supply and Training of Talent," OTA contractor report, 1987.

and articulation with admission standards; the integration of graduate education and research-oriented faculty into undergraduate curriculum; requirements for industrial experience and proven teaching ability for tenure; programs for gifted students; improved facilities; dropping shop and upgrading laboratories, retaining a 4-year curriculum but encouraging experimentation; a focus on design; a base curriculum of engineering science, not contemporary engineering practices; the inclusion of social and economic factors in solutions to technological problems; unification of analytical methods in all branches of engineering; and lifelong learning.

1956 Publication of the E. S. Burdell report (complementary to the Grinter Report), *General Education in Engineering—Report of the Commission for the Humanities: Social Research Project* (of the ASEE). Conclusions: more humanities and social sciences needed; rejected fears that this will either weaken engineering education or lead to superficial treatment of humanities and social sciences.

1959 Report to President Eisenhower by the President's Science Advisory Committee (Lee DuBridge, chairman), *Education for the Age of Science*, urged enhance the image of the teaching profession; improve high school education as preparation for science and engineering careers; reform curricula by unifying it along scientific principles common to engineering specializations, teach relation of engineering to social and governmental problems instead of parallel humanities/social sciences stem; promote the Ph.D. for engineers; provide special programs for gifted students; expand technical institutes; and retain faculty.

1966 Engineers Joint council response to interim ASEE *Goals of Engineering Education* Report: integrate teaching of engineering practice into its social context; focus on fundamentals, not current information; do not standardize curricula or accreditation; increase student-faculty interaction; promote lifetime learning; and expand the role of engineering professional societies in linking education to state-of-the-art practices.

1968 Publication of Final Report of the 5-year ASEE study, *Goals of Engineering Education*. It endorsed the Grinter Report on engineering science as the basis of engineering education. Recommendations: add 1 year of graduate study to basic engineering education; limit prerequisites and open the engineering major to transfers; expand cooperative and interdisciplinary programs; reduce credit hours for graduation; improve teaching of social and economic factors influencing, and influenced by, technology by integrating humanities and social sciences into the engineering curriculum; integrate research and undergraduate teaching; hire faculty with industrial experience, regardless of degrees; expand technical programs; and expand industry funding of engineering research; promote advanced engineering education (Ph.D.), continuing education, lifelong learning, professional registration by faculty. Predictions: M.S. will become the basic engineering degree; fewer programs/institutions; and the increasing use of engineering to solve social problems.

1968 Olmsted report for ASEE: integrate humanities and social sciences into 4-year programs; improve general education; retain humanities and social science faculty; and reduce the number of electives while retaining breadth.

1975 The Massachusetts Institute of Technology Center for Policy Alternatives Report (J. Herbert Holloman, chairman), *Future Directions for Engineering Education: System Response to a Changing World*, provoked by a "precipitous decline" in engineering enrollments and America's global dominance. It noted that engineering education was too responsive to "transient" changes. Recommended: prepare for declining enrollments; restore art of engineering to curriculum by teaching design; require work experience or cooperative education integrate humanities and social sciences into engineering curriculum; raise consciousness of "culture" of the sciences as opposed to their techniques; teach social, economic, political and legal constraints on engineering; expand 2- and 4-year technology programs; promote continuing education in engineering rather than management; expand evaluation; promote the engineering major as generic preprofessional training; and use industry more as a resource and sponsor.

1982 *The Quality of Engineering Education*, National Association of State University and Land-Grant Colleges (J. D. Kemper, chairman), cited problems of overenrollment, faculty shortages, and serious inadequacies in equipment, space, and facilities and recommended increased faculty salaries and industry support and government funding to upgrade the infrastructure.

1985 The National Academy of Engineering (NAE) published a 9-volume study, *Engineering Education and Practice in the United States,* chaired by J. A. Haddad.

1985 NAE report to the National Science Foundation (NSF), *New Directions for Engineering in the NSF* (Peter Likins, chairman).

1986 National Conference on Engineering Education, convened by the Accreditation Board for Engineering and Technology. Consensus recommendations: update undergraduate engineering education with mathematics concentration in probability, statistics, and numerical analysis; more breadth in basic sciences; expand humanities, social sciences, and communication skills; focus on design, including socioeconomic factors; intensify use of computers; introduce interdisciplinary coursework in real-world problem contexts; set admission standards that obviate need for remediation; strengthen faculty, requiring industrial experience and teaching effectiveness for tenure; continuing education; advisory committee of practicing engineers for each engineering education unit; raise fellowship stipends to one-half industry starting salary to attract U.S. graduate students; tighten the link of engineering education to engineering practice; encourage longer than 4-year curricula but do not mandate them; and increase role for engineers vis-a-vis executives, economists, and politicians in improving competitiveness.

1986 Final ASEE Report, *Quality in Engineering Education Programs* (W. Edward Lear, project director) cited problems of overenrollment, insufficient and obsolete laboratory equipment, and facilities shortage and deterioration. Recommended: re-emphasize production along with research; make industrial experience and effective teaching conditions of tenure; require test of spoken English for teaching assistants; institute structured continuing faculty education; implement computers and other new educational technologies; expand production of technicians; and improve laboratory teaching, assigning senior faculty to it.

1986 *The Quality of Engineering Education II*, follow-up to 1982 report (James E. A. John, chairman) recommended: promote U.S. citizen graduate study by raising fellowship stipends to one-half industry starting salary; fund large scale facilities improvement and maintenance; retain Ph.D. faculty with a healthy campus research environment; and produce more technicians.

1986 National Research Council, Office of Scientific and Engineering Personnel, *The Impact of Defense Spending on Nondefense Engineering Labor Markets.: A Report to the National Academy of Engineering*

1986 *Engineering College Research and Graduate Study: A 20-Year Statistical Analysis*, W. J. Fabricky, J. E. Osbourne and R. C. Woods.

1987 ASEE Report, *A National Action Agenda for Engineering Education* (E. E. David, chairman). Its eight recommendations: scale back the 4-year, necessarily limited curriculum to prepare for continuing education; make graduate education more practice-oriented; re-emphasize engineering design and manufacturing; improve undergraduate laboratories; attract more and better U.S. graduate students and faculty with higher salaries and research funding; bolster faculty development; support career-long education; and improve precollege mathematics and science education and introduction to engineering careers.

APPENDIX B

IEEE Engineering Skills Assessment Program

The purpose of the Engineering Skills Assessment Program (ESAP) is to provide guidance for members seeking to evaluate their skills in an electrical engineering field. ESAP is the logical starting point for professional development to help engineers advance their career objectives.

ESAP will address areas where:

- there is a rapidly changing technology,
- new career opportunities appear frequently,
- there is interest expressed by many employers,
- there is a readily identifiable knowledge base, and
- there are many recognized experts who can speak for the field.

In electrotechnology, careers are affected by rapid technological change, requiring engineers to update constantly what they have learned in college. Recognizing the need for professional development, the National Academy of Engineering, through its Committee on Career-Long Education for Engineers, stated in 1988 that professional growth and productivity are the responsibility of engineers. Accordingly, the IEEE Board of Directors has placed a high priority on developing member services that fulfill career-long education needs.

To help members avoid technical obsolescence and to take advantage of new career opportunities, the IEEE Educational Activities Board has developed the Engineering Skills Assessment Program (ESAP) that will:

- relate directly to a member's job and/or career objectives,
- concentrate on fast-moving technologies,
- provide personal feedback based on test results,
- be highly personalized, inexpensive and convenient,
- provide linkage across the engineering spectrum from research to manufacturing including the academic community,
- be valuable to both industrial and academic members,
- be developed and distributed through Societies, and
- not duplicate what is available elsewhere.

THREE COMPONENTS OF ESAP

Field Specific Knowledge Inventory (FSKI)

An inventory of knowledge elements which are deemed important by successful incumbents developed by volunteers. The FSKI is tested for validity against opinions of other successful practitioners. The final results are published in Society publications.

Self Assessment Test (TEST) and Answers

A multiple choice test is prepared and validated against the FSKI by standard statistical methods. The TEST and Answers are released in Society publications for members use at their discretion.

Guidance Information

References are given for each part of the FSKI and related TEST answers.

IMPLEMENTATION OF ESAP

ESAP, a new service to members, will yield a higher level of technical expertise through input from accomplished professionals. ESAP will have a teamwork approach between Educational Activities and Societies with responsibilities and costs assumed by each.

The participating Society will:

- identify appropriate technical fields for ESAP development
- identify successful incumbents for FSKI seminar panel and reviewer groups
- identify TEST writers
- identify TEST takers for test validation
- support volunteers in their preparation of ESAP components
- determine useful lifetime of the FSKI and TEST
- publish FSKI, TEST, GUIDANCE INFORMATION in Society publications
- promote programs to members.

The ESAP Committee will:

- provide staff support for Society ESAP Committees,
- guide FSKI, TEST process,
- support meeting costs for FSKI and TEST seminars,
- arrange for statistical consultations,
- provide evaluation of program,
- promote to Societies and Sections,
- support ESAP Committee expenses, and
- represent EAB/General Funds investment.

The ESAP project is primarily intended to facilitate the professional growth and development of IEEE members; however, there are potential organizational benefits to IEEE entities as well. The publication in society literature of the FSKI, TEST and GUIDANCE INFORMATION should attract new members. Member feedback should be useful to Societies and Sections in program planning and assessing future needs.

FOR MORE INFORMATION CONTACT:

IEEE
EDUCATIONAL ACTIVITIES
445 HOES LANE
P. O. Box 1331
PISCATAWAY, NJ 08855-1331
(201) 562-5489

August 1989

EVIDENCE OF ADAPTABILITY IN THE LABOR MARKET FOR ENGINEERS: A REVIEW OF RECENT STUDIES

Robert C. Dauffenbach
Oklahoma State University
and
Michael G. Finn
National Research Council

Introduction

Understanding the degree of adaptability in the labor force is important for analysis of many important issues that need to be addressed today in planning for the needs of tomorrow. In the occupational domain of scientists and engineers, that need for understanding is paramount. It is arguable that no other occupational domain is more crucial to the present and future competitiveness of the U.S. economy.

Adaptability can be measured in a variety of ways, such as by the ability of individuals trained in given disciplines to attain career success in alternative occupations; by the rates of mobility among firms, occupations, and regions of the country; and by the time and resource costs associated with retraining. The willingness and ability of individuals trained in one field of study but working in alternative occupations would seem to be a primary measure of flexibility in the economy. In this sense, anyway, the extent of flexibility among science- and engineering-educated personnel would seem quite large. The Survey of Income and Program Participation (SIPP) provides national labor force evidence on education/occupation correspondence. While this evidence is highly aggregated and is, given the relative small sample base for the SIPP surveys, subject to large sample variation, it is practically the only evidence available. The education and occupation correspondence shares for bachelor's and higher degree holders who are employed in scientific and engineering (S&E) occupations are as follows:

S&E Major Degree Field	Yield to S&E Employment
Engineering/computing	46.3%
Math/statistics	24.9%
Agriculture/forestry	15.4%
Biology	19.7%
Physical/earth science	34.7%
Psychology	17.1%
Economics	15.7%
Social sciences	5.9%

The largest of these numbers, 46.3 percent for engineers, is surprisingly small. Note that this number reflects all S&E employment, not just simply engineering employment (which, of course, dominates this category). The next highest value is for physical and earth sciences, at slightly more than one-third. Social science discipline yields are particularly low. Evidently, most individuals trained as scientists and engineers work in jobs outside science and engineering. Unfortunately, there is no knowledge of the extent to which such individuals are willing and able to undertake S&E jobs—that is, the rate of back flows. It is apparent, however, that among S&E graduates the yield to S&E employment is quite low.

In a study by Dauffenbach (1989), an analysis of various adaptability issues in science and engineering labor markets is presented. Analysis of the correspondence between occupation and education is seen in his study as a primary vehicle for appraisal of flexibility in labor markets. Dauffenbach provides several cross-tabulations of detailed occupation by field of study for science and engineering occupations using NSF's 1982 postcensal survey data. A major finding is that while detailed field of study is a good predictor of occupational pursuit, the amount of variance is surprisingly high.

Such prevalence of non-exact correspondence between occupation and education can be taken as evidence of flexibility in the S&E labor market. Dauffenbach hypothesizes, however, that such flexibility is not without attendant costs in the form of diminished productivity. Presumably, if there are productivity differences between workers who are appropriately credentialed and others doing the same job, these differences should show up, systematically, in salaries. Thus, he undertakes an extensive analysis of salary differentials by applying the statistical methodology of multiple regression analysis to the aforementioned NSF survey data. These differentials, in general, support the notion that quality differentials result from non-alignment of degree field and occupational pursuit. In addition, because of the variety of other factors that need to be held constant in order to have an unbiased assessment of the impact of education/occupation correspondence on salaries, the regression results provide a general assessment of the various factors on salary differentials, such as race, sex, primary work activity, industry, occupation, and professional work experience. Also, because in the 1982 postcensal survey, certain questions pertaining to mobility were asked, it is possible to investigate the impact of inter-firm and inter-occupational changes on earnings.

Degree and Employment Fields

Dauffenbach provides separate regression results for each of the major domains of S&E employment: engineering, biological sciences, math/computer science, physical sciences, and social sciences. He concentrates on detailed categories of field of study and occupation in his analysis and explores the correspondence between field of study and occupation on three levels.

First, there is the exact match level in which an individual is working in an occupation that corresponds exactly to his or her field of study, such as a mechanical engineer holding a highest degree from the mechanical engineering field. Second, there is the associated-field level of correspondence, such as a mechanical engineer working as an aeronautical engineer. Third, there is the non-associated level, such as an individual with a highest degree in education working as an engineer. Of course, there could well be differences among major degree fields. Physical science graduates would be expected to be more readily interchanged for engineers than, say, business school graduates. Consequently, non-associated fields were divided into several major fields of study (including health, education, business, and "all other") in addition to the major science fields. After substantial investigation it was decided to use 35 distinct degree fields, which were mapped into 40 S&E occupations. These categories represent the numerically significant fields and occupations in the NSF postcensal survey.

In the five sets of regressions provided by Dauffenbach, a basic point of interest is the extent to which individuals appear to be working in fields not directly associated with their detailed major field of study (Table 1). These results are interpreted in the following

Table 1. Degree Field Shares, by Field of Employment (in percent)

Field of Study	Occupational Field of Employment				
	Engineering	Math & Computer	Physical Science	Biological Science	Social Science
Exact Match	54.9	21.6	71.1	61.3	68.4
Engineering	25.1	11.0	4.8	0.8	1.0
Math & Computer	3.0	23.0	1.9	0.3	1.2
Physical Science	5.0	5.2	8.2	4.2	0.4
Biological Science	1.2	3.4	7.4	26.1	1.3
Social Science	1.3	8.4	1.9	1.5	10.5
Education	1.5	5.6	1.5	2.4	3.2
Health	0.1	0.4	0.6	0.5	0.3
Business	4.6	13.0	0.9	0.5	4.9
All Other	3.3	8.4	1.7	2.4	8.8

manner. Among all of the nearly 20,000 observations of employed engineers, about 55 percent had an exact match between their detailed employment field and the detailed field of their highest degree earned. Another 25.1 percent of the employed engineers had an engineering degree, but their degree field did not match their employment field. This leaves a residual of about 20 percent with a degree in a non-associated field. For engineering, the most prominent of these non-associated fields was physical science, followed closely by business. Other results are read in a similar manner. Note also that individuals with their highest degree in engineering represent sizable proportions of both the math and computer science and the physical science employment fields.

Earnings Differentials

Of primary interest is the impact on earnings differentials associated with not having an exact match or having a highest degree in a non-associated field of study—if, in fact, there are real productivity differentials associated with individuals who do not match in

Table 2. Regression Estimates of Earnings Differentials (in percent)

Field of Study	Occupational Field of Employment: Engineering	Math & Computer	Physical Science	Biological Science	Social Science
Exact Match	10.04*	7.27*	16.42*	0.67	9.19*
Engineering	9.87*	7.76*	18.58*	15.72*	16.84*
Math & Computer	9.93*	8.51*	21.61*	6.35	10.62
Physical Science	5.64*	5.97*	12.04*	8.93	18.24
Biological Science	0.81	-0.21	5.79	-1.50	-12.46
Social Science	1.83	-0.19	0.78	1.48	1.74
Education	-5.03*	-2.67	4.90	-4.26	-8.91
Health	3.91	3.81	12.25	9.40	-59.86*
Business	3.81*	2.83	23.20*	16.51	6.66
All Other	-	-	-	-	-

* The coefficient is statistically significant at conventional levels.

terms of degree field and employment field (Table 2). The coefficients are read relative to the salary associated with the "all other" fields of study, which is the excluded group in the regression analysis. Thus, we see that an exact match in the engineering employment field pays about a 10 percent differential above those who have a degree in the "all other" field. However, having a non-exact match but still having an engineering degree pays almost as much, a 9.87 percent differential. Note also that having a degree in math/computer science and working as an engineer also pays a handsome differential of 9.93 percent. Yet, having a highest degree in biology, physical science, social science, education, health, or business pays somewhat less.

In the math/computer domain, the coefficient for the associated field (same general field, but not an exact match) is actually higher than the exact-match coefficient. Engineering graduates earn about the same; physical science graduates, slightly less. Other fields of study are somewhat lower. Business graduates are a large contributor to this employment field. They earn 4-5 percent less than engineers working as math/computer specialists.

A total of 75 percent of those working in the physical sciences have physical science degrees. In this employment domain, individuals with biological science degrees are the most frequent other contributor. They earn significantly less, about 7-11 percent. Engineers, another significant contributor, earn about the same amount, if not a little higher than the exact-match.

Also of interest in these results is the finding that in all of the major S&E employment fields, those who have engineering degrees earn as much or more than those who have exact matches with their employment field. This result seems especially significant in regard to flexibility of engineers. However, as noted previously, engineers represent a sizable proportion of only the math/computer and physical science employment categories. Still, the biological and social sciences are large employment fields and even a 1 percent composition of engineers is not an insignificant number.

Mobility

Dauffenbach's results also provide information on the extent of mobility of various types: change in employer, in occupation, and in responsibilities. Table 3 provides these gross mobility rates for the major S&E employment fields, 1976-1982. Change in responsibilities appears to be rather frequent among the S&E categories and about 1 in 4 or

Table 3. Estimates of Mobility Rates, 1976-1982 (in percent)

	Occupational Field of Employment				
Type of Change	Engineering	Math & Computer	Physical Science	Biological Science	Social Science
Employer	21.99	27.50	23.22	19.34	23.66
Occupation	10.65	16.99	11.32	10.13	12.16
Responsibilities	32.32	36.15	28.63	28.51	26.29

5 changed employer within the six years. Occupational mobility is substantially lower. Those working in the math/computer employment field were the most likely to be mobile occupationally, about 17 percent as compared to the more common 10-12 percent. These mobility figures are derived from retrospective questions—that is, in the 1982 survey, respondents were asked how their jobs had changed since 1976. Mobility results tend to be substantially higher when tabulated from longitudinal data, a result that is most likely a consequence of coding error.

These findings from Dauffenbach's study allow one to place some bounds on the extent of mobility, its character, and earnings consequences. The results imply that there is a fairly high degree of adaptability among engineers, math/computer specialists, and physical scientists. This finding is validated by (1) the magnitudes of individuals having their highest degree in one of these fields, but working in one of these other fields, and (2) the essentially nil pay differentials among these degree fields within each respective employment field. Other fields, especially business disciplines, contribute significantly at time, but in general have substantially lower pay differentials. When these results are coupled with the evidence that the majority of S&E degree holders do not work anywhere in science or engineering, the extent of flexibility is large, indeed. A major gap in our knowledge is the extent to which such individuals are both willing and able to return to S&E career pursuits. This is a knowledge gap in great need of being closed.

Other Findings

Other recent studies find similar results to those of Dauffenbach and also examine explicitly the value of an engineering degree for persons in management jobs. Korb (1987) studied the employment of 1983-84 baccalaureate degree recipients in 1985, using a Department of Education survey of recent graduates that included all degree fields. Not surprisingly she found that engineering graduates reported the highest salaries and that those with nonengineering jobs 1-2 years after graduation earned somewhat less than those with engineering jobs. However, the engineering degree seems to be valued more highly for the principal nonengineering jobs entered by recent engineering graduates when compared with the earnings of other college majors in the same jobs. For example, she found that engineers were employed as technicians after graduation much less frequently than were biological sciences graduates (9 percent versus 40 percent), but that they reported much higher earnings than the biological sciences graduates when they did work as technicians ($22,000 versus $15,000). Only 5 percent of engineering graduates took jobs as managers 1-2 years after graduation. They earned less than the engineering graduates who took engineering jobs but more than business/management graduates who took jobs as managers. In short, the Department of Education survey indicates that new engineers are valued most highly for engineering jobs, but they are valued more highly than other majors in all of the occupations they enter. This suggests a great deal of adaptability.

The strong labor market for recent engineering graduates is widely known, but are persons trained as engineers as highly valued later in their careers? Evidence from another Department of Education survey suggests that the answer to this question is "yes." James et al. (1989) examined data from the National Longitudinal Study of the High School Class of 1972, and used data they reported about their jobs in 1986, about 9-10 years after the typical person in the sample had completed a bachelor's degree. Their study was innovative in that it controlled for differences in ability as measured by SAT scores, for differences in college grades, and for the number of math courses taken in college, all factors that enhance earnings. Still, engineering graduates were found to have the highest earnings in 1986, about 20 percent higher than the second highest-paid bachelor's degree field, business. Relevant to the adaptability question is the fact that engineering graduates who held nonengineering jobs earned salaries that were just as high as those with engineering jobs. On the other hand, business majors seem less versatile: they reported high wages only when they held jobs as managers.

We could cite other studies which document that engineers earn more when they devote a relatively high percentage of their time to management activities (e.g., Finn, 1985). However, this is probably widely accepted and needs little discussion.

Summary

The studies cited here indicate that persons with engineering degrees do frequently take jobs outside engineering—not only in the physical, mathematical, and computer sciences, but also as managers and in other capacities such as sales. Salary is a simple and imperfect summary measure of the adequacy with which they perform these jobs. Yet findings of these studies of earnings are unambiguous. They indicate that, no matter what the occupation studied to date, persons trained as engineers do at least as well as persons trained in any other degree field. This might be the result not merely of the adaptability provided by an engineering education, but also of the superior ability or willingness to work hard that characterizes the students who complete engineering degrees. However, we note that one study that was able to control for some measures correlated with native ability and willingness to work hard (SAT scores, college grades) found evidence of the same high degree of adaptability as did the other studies without such measures.

It would be incorrect, however, to infer that degree field doesn't matter, that all college graduates are adaptable. The Dauffenbach study indicates clearly that persons without engineering degrees earn less than those with engineering degrees when the job is engineering. The one exception seems to be the math and computer science degree-recipients. They seem to show as much adaptability as engineering degree-recipients, at least for jobs in the broad groupings of engineering, mathematical sciences, computer science, and physical science.

References

Dauffenbach, Robert C. 1989. *The Issue of Quality in the Market for Scientists and Engineers - A Report to the National Science Foundation.* Stillwater, Okla.: Oklahoma State University.

Finn, Michael G. 1985. *Foreign National Scientists and Engineers in the U.S. Labor Force, 1972-1982* (ORAU-244). Oak Ridge, Tenn.: Oak Ridge Associated Universities.

James, Estelle, N. Alsalam, J. C. Conaty, and D. To. 1989. College quality and future earnings: Where should you send your children to college? *American Economic Review* 79(May):247-252.

Korb, Roslyn A. 1987. *Occupational and Educational Consequences of a Baccalaureate Degree.* Washington, D.C.: U.S. Government Printing Office.

ADAPTABILITY OF THE ENGINEERING WORK FORCE: INFORMATION AVAILABLE FROM THE BUREAU OF LABOR STATISTICS

Alan Eck
Bureau of Labor Statistics

Introduction

Through its Occupational Outlook Program, the Bureau of Labor Statistics (BLS) provides current occupational information to a wide variety of users. This program, conducted by the BLS' Office of Employment Projections, has seen its primary audience change from the unemployed of the 1930s depression, to veterans returning to the civilian job market after World War II, and since the late 1940s, to high school students seeking assistance in choosing a career. Throughout, however, its focus has remained the same: to provide current information about job duties and working conditions, current employment, training requirements, earnings, and the outlook for jobs in about 200 occupations. The program's major publication, the *Occupational Outlook Handbook*, appears biannually and is widely used in high school vocational education programs.[1]

To best serve its audience, the *Handbook* utilizes a nontechnical, narrative format in presenting information. However, the information is based on careful analysis of BLS and other data. Indeed, the program resides in the BLS to facilitate access to extensive survey data available in other program offices. To provide students, counselors, researchers, training program planners and others interested in the specific data used to develop information in the *Handbook*, a statistical supplement, *Occupational Projections and Training Needs*, also is produced biannually.[2]

---

[1] *Occupational Outlook Handbook, 1988-89 Edition* (Bulletin 2300), Washington, D.C.: Bureau of Labor Statistics, April 1988, is the most recent edition. The next is scheduled for publication in April 1990.

[2] *Occupational Projections and Training Data, 1988 Edition* (Bulletin 2301), Washington, D.C.: Bureau of Labor Statistics, April 1988.

Unlike the Center for Education Statistics, which concentrates on analyzing information about teachers, or the National Science Foundation (NSF) on scientists and engineers, occupational analysis within the Office of Employment Projections embraces all occupations. Analysis of engineering employment is part, but certainly not all, of this effort.

Thus, to support the Occupational Outlook Program and for other purposes, the Bureau develops information about all occupations. A detailed system for projecting the labor force, gross national product, productivity, and other economic variables is the cornerstone for comprehensive occupational employment projections.[3] Projections are valuable because they permit calculation of expected and absolute rates of change for about 600 occupations. The former provides a partial measure of job opportunities; the latter is useful for identifying the rapidly growing occupations that are generally considered to provide better employment prospects.

While occupational employment projections are valuable, they are only a portion of desirable occupational data, and the BLS supports efforts to develop additional information. For example, since growth is an incomplete measure of job opportunities and since job opportunities resulting from the need to replace workers who leave an occupation greatly exceed those due to growth, the potential for developing replacement needs data was explored. Also, the desirability of providing information identifying the characteristics of entrants and the existence of career ladders to students and training program planners was recognized, and research to develop appropriate information was undertaken. These research efforts utilized existing BLS data and provide insights into the adaptability of engineers.

In this paper, the two major BLS occupational data sources are discussed, and a brief overview of the information about engineers is presented. Be forewarned, however, that this paper does not provide an intensive analysis of the labor market for engineers.

Data Sources

Current and projected employment data appearing in the *Occupational Outlook Handbook* and Occupational Projections and Training Needs Data utilize information from

[3] A series of articles in the September 1987 *Monthly Labor Review* present the results. The articles and a description of the projection methodology appear in Projections 2000, Bulletin 2302 (Bureau of Labor Statistics, March 1988).

the BLS national industry/occupation matrix. Basic data for that matrix come from Occupational Employment Statistics (OES) surveys conducted periodically by State employment security agencies under a BLS-State cooperative program.[4] The OES survey obtains information from establishments rather than individuals. Forms containing occupational descriptions are provided employers, who then classify their employees by occupation. Appendix 1 presents OES survey definitions for engineers. None of these definitions require an individual to possess a college degree in order to be classified as an engineer. Engineering teachers are not included among engineers; they are included in the data for college and university faculty. Also, sales engineers are included with sales workers rather than engineers. The OES survey obtains only wage and salary employment data and provides no demographic information about employees. Because data are obtained from employers, the results count jobs, not individuals. For example, an individual with two jobs would be included in the data twice, once with each employer.

To complete the national industry/occupation matrix, OES survey data are combined with decennial Census of Population and Current Population Survey (CPS) occupational employment information for the agriculture and private household industries as well as for self-employed and unpaid family workers. Office of Personnel Management data provide occupational employment information for the federal government.

Although BLS uses OES survey employment estimates in its Occupational Outlook Program, the survey's usefulness in examining the adaptability of engineers is limited since no demographic data are available. However, other BLS data—from the CPS—do provide information about the adaptability of engineers.[5] The CPS is a monthly survey of approximately 55,000 households conducted by the Bureau of the Census for the BLS. Interviewers conduct the survey and ask questions to determine if individuals are employed, unemployed, or out of the labor force. Information about industry and occupation of employment—as well as about age, sex, education, and many other characteristics—also is obtained. After interviewers return the forms, Bureau of the Census clerical personnel code responses to the survey questions—"What kind of work was ____ doing? (For example: electrical engineer, stock clerk, typist, farmer.)" and "What were ____'s most important activities or duties? (For example: types, keeps account

---

[4]Additional information about the Occupation Employment Statistics survey appears in the publication *BLS Handbook of Methods*, Bulletin 2285, (Bureau of Labor Statistics, April 1988), Chapter 3.

[5]A description of the CPS appears each month in "Explanatory Notes," Household Data section, of the BLS publication, *Employment and Earnings.*

books, files, sells cars, operates printing press, finishes concrete.)"—into 1 of about 500 occupations established for use with the decennial Census of Population occupational classification system. Since the classification system changes after each decennial Census, occupational data may be consistent only for about 10 years. For example, from January 1972 through December 1982, the CPS used the 1970 Census of Population occupational classification system; since January 1983, the 1980 system has been used. Fortunately, changes in the system do not significantly impact the data about engineers. The job titles included in each of the engineering occupations from the *1980 Census of Population Classified Index of Industries and Occupations* are presented in Appendix B.[6] Engineering teachers are included in the data for college and university faculty, and sales engineers are now included with sales workers.

Unlike the OES survey, in the CPS responses of individuals rather than employers determine occupation. There are no education standards; individuals with less than four years of college may report and be classified as an engineer. In some cases this may result in occupational upgrading (technicians may describe themselves as engineers). On the other hand, individuals without degrees may correctly describe themselves as engineers. An additional result of the survey design is that the employment data count individuals, not jobs. For this reason and others, occupational employment data from the CPS differ from current and projected employment data appearing in the national industry/occupation matrix.

Table 1 compares 1986 CPS, OES, and (NSF) survey employment data for engineers. While a discussion of how NSF data are derived and why they differ from the OES and the CPS is beyond the scope of this paper, these data provide a comparison with another major data source. Suffice to say that the NSF estimate of 2.6 million engineers, which includes college faculty, is much higher than the 1.7 million CPS estimate and the 1.4 million OES survey-based estimate.

Information from the Current Population Survey

Many types of information provide insights into the adaptability of engineers and the labor market environment that requires adaptation; the following reviews those available

[6] *1980 Census of Population: Classified Index of Industries and Occupations* (PHC80-R4), Washington, D.C.: Bureau of the Census, November 1982.

Table 1. Comparison of 1986 Employment for Engineers (in thousands)

Current Population Survey Title	Current Population Survey[a]	National Industry-Occupation Matrix[b]	National Science Foundation[c]
Engineers	1,749	1,371	2,561
Aerospace engineers	93	53	112
Metallurgical & materials engineers	26	18	59
Mining engineers	9	5	19
Petroleum engineers	32	22	38
Chemical engineers	59	53	163
Nuclear engineers	10	14	25
Civil engineers	233	199	366
Agricultural engineers	3	-	-
Electrical & electronic engineers	550	401	581
Industrial engineers	203	117	151
Mechanical engineers	287	233	514
Marine engineers & naval architects	13	-	-
Engineers, n.e.c.	228	257	532

[a] Based on tabulation of Current Population Survey micro-data for all months in 1986.

[b] George T. Silvestri and John M. Lukasiewicz, "A look at occupational employment trends to the year 2000," *Monthly Labor Review*, September 1987, Table 3, pg. 49-54.

[c] National Science Foundation, "Profiles - Mechanical Engineering: Human Resources and Funding" (NSF 87-310), (Washington, D.C.) Table 1.

- Data not available.

from the CPS. First, selected demographic characteristics are examined to obtain a sense of how engineers differ from other professional workers and from all employees. Then, employment trend data assess changes in demand for their services and changes in the characteristics of workers. Finally, data about movements describe the sources of entrants and the destination of leavers. Because significant errors can exist in data for small occupations, information is presented only for engineering occupations with 100,000 or more employees in 1988.

Selected Characteristics

When compared to all workers (Table 2), engineers are older than all workers (median years of age: 38.8 versus 36.2), have more education (median grade of school

Table 2. Selected Characteristics of Workers, 1988

Occupation	1988 Employment (thousands)	Median Age	Median School Grade Completed	Percent of 1988 Employees on Part-Time Schedules	Females	White
Total employed, age 16 & over	115,003	36.2	12.9	17.2	45.0	86.8
Professional specialty occupations	15,010	38.5	17.2	15.1	49.8	89.3
Engineers	1,815	38.8	16.8	2.2	7.4	89.6
Aerospace	117	41.7	17.1	1.9	6.8	87.4
Civil	224	39.2	16.9	2.2	6.3	89.4
Electrical & electronic	572	37.9	16.9	1.7	8.1	88.1
Industrial	220	39.4	16.3	1.5	13.0	90.5
Mechanical	300	39.5	16.8	1.7	3.7	91.3
Engineers,n.e.c.	228	39.7	16.8	3.5	6.8	90.5
Registered Nurses	1,561	37.4	16.0	26.1	94.7	86.1
Teachers, elementary	1,423	40.1	17.4	10.7	84.6	88.0
Teachers, secondary	1,193	41.1	17.7	9.5	51.2	90.4
Engineering & related technologists & technicians	934	34.3	13.9	5.6	18.8	89.5
Electrical & electronic technicians	326	34.1	13.8	3.7	14.5	87.4
Engineering technicians, n.e.c.	224	34.4	14.0	9.4	31.1	87.9
Drafting occupations	289	34.0	14.1	4.9	15.8	90.9

SOURCE: 1988 annual average Current Population Survey data.

completed: 16.8 versus 12.9), have much smaller proportions of part-time workers (2.2 versus 17.2 percent) and female workers (7.4 versus 45.0 percent), and have a higher proportion of white workers (89.6 versus 86.8 percent). Except for smaller proportions of part-time and female workers, however, the characteristics of engineers and all professional workers are very similar.

Table 3 complements the information on educational attainment provided in Table 2 by distributing engineers by highest grade completed. Seventy-two percent completed 16 or more years of school and probably have a college degree.

CPS data also provide a glimpse at the relative stability of engineers. Occupational tenure data measure the length of time individuals have done the kind of work they are now doing while working for either their current or any previous employer. The median years of tenure in their current occupation was 10.5 for engineers, slightly higher than the 9.6 years for all professional workers but significantly higher than the 6.6 years for all employed workers (Table 4). Another measure that engineers tend to stay in the occupation is provided by the proportion with 20 or more years tenure in the occupation. At 28.2 percent, the proportion is much higher than the 20.0 percent for all professional workers and 14.6 percent for all workers.

Employment Trends

Demand for engineers has grown significantly over the last 25 years. Employment increased from 985 thousand in 1963 to 1.805 million in 1988, an 83 percent increase (Table 5). Only during 1968-1973, when curtailments in the space program and military involvement in Vietnam sharply reduced the need for engineers, was there any significant deviation in the upward trend. Somewhat surprisingly, the 73 percent increase is about the same as the 66 percent increase in total employment. In the last decade, however, employment of aerospace, and electronic engineers has grown at least twice the rate as total employment (Table 6).

Table 3. Engineers, by Highest Grade of School Completed, 1988

Occupation	Percent of 1988 Employment			
	Total	16 years or more	13-15 years	12 years or less
Engineers	100.0	72.3	16.4	11.3
Aerospace	100.0	81.8	12.7	5.5
Civil	100.0	75.0	13.5	11.5
Electrical & electronic	100.0	73.5	16.6	9.9
Industrial	100.0	56.5	22.9	20.6
Mechanical	100.0	71.0	17.6	11.4
Engineers, n.e.c.	100.0	72.1	19.1	8.8

SOURCE: 1988 annual average Current Population Survey data.

Table 4. Years of Tenure in Occupation, 1988

Occupation	Total Employed (thousands)	Median	Percent of Employees: Total	3 or less	4-9	10-19	20 or more
		Years of Tenure in Current Occupation					
Total employed, age 16 & over	109,090	6.6	100.0	36.5	26.1	22.9	14.6
Professional specialty occupations	14,448	9.6	100.0	23.7	27.5	28.9	20.0
Engineers	1,784	10.5	100.0	19.9	27.6	24.4	28.2
Aerospace	109	9.6	100.0	22.8	28.3	10.3	38.7
Civil	237	13.0	100.0	17.8	19.7	28.9	33.5
Electrical & electronic	520	10.4	100.0	18.2	29.6	25.3	26.9
Industrial	213	8.9	100.0	24.5	29.0	24.6	21.9
Mechanical	288	11.4	100.0	21.9	23.7	24.2	30.2
Engineers, n.e.c.	269	10.0	100.0	19.8	30.0	25.2	24.6
Registered Nurses	1,538	9.3	100.0	20.7	31.5	28.0	19.8
Teachers, elementary	1,412	12.4	100.0	14.3	24.3	41.3	20.1
Teachers, secondary	1,182	12.5	100.0	14.9	22.3	39.4	23.4
Engineering & related technologists & technicians	847	7.6	100.0	29.7	29.4	24.1	16.8
Electrical & electronic	294	6.9	100.0	33.7	26.6	25.7	14.0
Engineering technicians, n.e.c.	205	7.7	100.0	28.2	30.5	29.5	11.7
Drafting occupations	281	8.0	100.0	28.8	29.9	19.1	22.2

SOURCE: January 1987 Current Population Survey.

Table 5. Employment, Selected Years (in thousands)

Occupation	1963	1968	1973	1978	1983	1988
Total employed, age 16 & over	69,084	75,920	84,409	94,375	100,832	114,968
Engineers	985	1,193	1,094	1,265	1,572	1,805
Aerospace	63	76	59	59	80	115
Civil	158	167	156	160	211	218
Electrical & electronic	255	290	272	329	450	573
Industrial	113	149	167	206	210	221
Mechanical	201	227	178	216	259	297
Engineers, n.e.c.	103	133	146	173	192	230

SOURCE: Annual average Current Population Survey data.

Table 6. Percent Change in Employment, Selected Years

Occupation	Percent Change 1963-68	1968-73	1973-78	1978-83	1983-88
Total employed, age 16 & over	9.9	11.2	11.8	6.8	14.0
Engineers	21.1	-8.3	15.6	24.3	14.8
Aerospace	20.6	-22.4	.0	35.6	43.8
Civil	5.1	-6.6	2.6	31.9	3.3
Electrical & electronic	13.7	-6.2	21.0	36.8	27.3
Industrial	12.0	12.1	23.4	1.9	5.2
Mechanical	12.9	-21.6	21.4	20.0	14.7
Engineers, n.e.c.	29.1	9.8	18.5	11.0	19.8

SOURCE: Annual average Current Population Survey data.

As the data presented earlier indicated, engineers are predominantly white males. An increase in the proportions of non-whites and females over time indicates engineers are expanding their traditional source of entrants. Such appears to be the case. Since 1963 the proportion of non-whites has increased from 2.1 to 10.2 percent, and the proportion of females from 0.7 to 7.3 percent (Table 7).

Gross Movements

Employment opportunities result from the creation of new jobs and the need to replace workers. "Gross separation" data identify those workers who leave an occupation and who must be replaced if employment levels are to be maintained. Information about gross separations not only identifies employment opportunities, but also indicates the relative attachment of individuals to an occupation.

Using the CPS as a data base, a methodology has been developed to estimate the proportion of workers leaving an occupation.[7] Briefly, at 1-year intervals, 50 percent of the households in the CPS sample are the same. Individuals who had not changed residence were identified in each survey by matching data on computer tapes about the household address and information about the age, sex, and race of the individuals. A matched sample for each of 12 months was created and data describing changes in labor

[7]The methodology is described in detail in *Occupational Projections and Training Data, 1982 Edition* (Bulletin 2202), Washington, D.C.: Bureau of Labor Statistics, December 1982, Appendix B.

Table 7. Non-whites and Females, Selected Years, in percent

Occupation	1963	1968	1973	1978	1983	1988
Total employed, age 16 & over						
Non-whites	10.3	10.8	10.8	11.2	11.8	13.2
Females	25.6	36.6	38.4	41.2	43.7	45.0
Engineers						
Non-whites	2.1	3.3	3.7	5.5	8.1	10.2
Females	.7	.7	1.3	2.8	5.8	7.3

SOURCE: Annual average Current Population Survey data.

force status tabulated. Matched data about changes in the labor force then were merged with data on occupational changes from a special study conducted as part of the January 1987 CPS. The results, termed "merged data," provide a composite description of movements into, out of, and between occupations over a 1-year period: they measure gross movements.

Table 8 presents 1986-87 gross separation data for engineers. Overall, 8.4 percent left the detailed occupation: about half, 4.3 percent, transferred to another occupation while the remainder became unemployed (1.5 percent) or left the labor force (2.6 percent).[8] Merged data show relatively few engineers leave from one year to the next. Their rate was slightly lower than the 10.8 percent for all professional workers and much lower than the 18.0 percent for all employed persons. The merged data also identify differences in separation rates among engineering specialities. Separation rates hovering around 6 percent are observed for the most technical groups—aerospace, civil, electrical and electronic, and mechanical—about one-half the rate for industrial and "engineers n.e.c." Over time, the levels of the rates and differences between occupations have remained relatively constant (Table 9). Industrial engineers consistently exhibit the highest separation rate.

Merged CPS data also can be tabulated to provide information about entrants. This group includes individuals who entered the occupation to fill newly created jobs as well as to replace engineers who left. The results reveal 5.8 percent of engineers were not in the occupation a year earlier. The largest group consists of individuals transferring from other

[8]Transfers measure changes between detailed occupations and include employment in a different engineering specialty.

Table 8. Separation Rates for Selected Occupations (in percent)

Occupation	Separation Rates, 1986-87			Not in the Labor Force				
	Total	Transfers to Other Occupations	Unemployed	Total	Household Responsi-bilities	School	Disabled	Other, including Retired
Total employed, age 16 & over	18.0	8.4	2.7	6.9	2.7	1.2	0.2	2.8
Professional specialty occupations	10.8	4.7	1.1	5.0	2.3	.8	.1	1.8
Engineers	8.4	4.3	1.5	2.6	.3	.3	.1	1.9
Aerospace	4.6	2.5	.7	1.4	.0	.0	.0	1.4
Civil	6.6	1.8	2.1	2.7	.0	.3	.2	2.3
Electrical & electronic	7.1	3.1	1.6	2.5	.2	.5	.1	1.8
Industrial	11.5	7.8	.9	2.7	.4	.0	.0	2.4
Mechanical	4.6	1.9	1.2	1.5	.0	.0	.2	1.4
Engineers, n.e.c.	14.2	8.4	2.1	3.7	.6	.4	.0	2.7
Registered Nurses	7.6	2.1	.5	5.0	3.4	.4	.1	1.1
Teachers, elementary	11.1	4.7	.6	5.9	3.4	.6	.1	1.7
Teachers, secondary	9.1	4.6	.5	4.0	1.9	.3	.0	1.8
Engineering & related technologists & technicians	11.5	6.2	2.4	2.9	.2	1.0	.1	1.6
Electrical & electronic technicians	10.5	6.2	2.5	1.8	.0	.6	.0	1.2
Engineering technicians, n.e.c.	14.3	6.0	2.5	5.7	.7	2.8	.5	1.7
Drafting occupations	12.3	7.2	2.6	2.4	.0	.3	.0	2.0

SOURCE: Merged 1986-87 Current Population Survey data.

Table 9. Separation Rates for Engineers, Selected Years (in percent)

Occupation	Separation Rates			
	Total	Transfers to Other Occupations	Unemployed	Not in the Labor Force
Engineers				
1977-78	8.1	5.1	0.7	2.3
1980-81	7.4	4.4	1.0	2.0
1983-84	8.6	4.4	1.1	3.0
1986-87	8.4	4.3	1.5	2.6
Aerospace engineers				
1977-78	8.0	7.2	.4	.4
1980-81	3.4	3.0	.4	.0
1983-84	4.4	.0	.0	4.4
1986-87	4.6	2.5	.7	1.4
Civil engineers				
1977-78	4.8	1.4	.8	2.6
1980-81	6.9	4.3	.6	2.0
1983-84	4.9	2.7	.9	1.3
1986-87	6.6	1.8	2.1	2.7
Electrical & electronic engineers				
1977-78	7.1	4.4	.6	2.1
1980-81	4.0	1.6	.7	1.7
1983-84	5.1	2.6	.5	2.0
1986-87	7.2	3.1	1.6	2.5
Industrial engineers				
1977-78	11.4	7.6	1.1	2.7
1980-81	14.7	9.5	2.2	3.0
1983-84	15.3	9.4	1.7	4.2
1986-87	11.4	7.8	.9	2.7
Mechanical engineers				
1977-78	4.6	1.7	.6	2.3
1980-81	6.3	4.4	.5	1.4
1983-84	8.5	1.8	1.9	4.8
1986-87	4.6	1.9	1.2	1.5
Engineers, n.e.c.				
1977-78	7.6	4.4	.7	2.5
1980-81	4.8	1.2	1.0	2.6
1983-84	11.3	6.6	1.3	3.4
1986-87	14.2	8.4	2.1	3.7

SOURCE: Merged Current Population Survey data.

occupations (3.6 percent). Of those not working, the largest group was out of the labor force and attending school (0.8 percent).

Unfortunately, merged data on entrants suffer from the limitation that individuals who change residences between surveys are excluded. This limitation may be especially significant in occupations such as engineering where new college graduates comprise a large portion of entrants. Merged data for such occupations understate entrants because many individuals move after completing school and thus are not counted in the second survey even though they are likely to be working as an engineer. While the merged data indicate a lower percentage entering than leaving (8.4 versus 5.8 percent), much of the difference is attributable to the observed decline in engineering employment; the difference is not solely the result of errors in the data.

Another limitation of the data is the inability to identify entrants who recently completed school. Because individuals are identified by labor force status, entrants who are students and who are working cannot be distinguished from qualified engineers: new graduates may be included in data for persons transferring into engineering.

While based on a sample about one-tenth the size of the merged data, there is another source of CPS data about entrants. A special supplement to the January 1987 CPS asked employed engineers and others if they were employed in January 1986 and, if so, in what occupation. These data estimate that 111,000 individuals—6.2 percent of all engineers—have entered the occupation since January 1986. The larger group—3.4 percent—transferred from another occupation while the remaining 2.8 percent had not been working the previous year. These results are similar to those obtained from the merged data (Table 10).

Data from the January 1987 survey also provide information about engineers who changed employers, but not occupation. These data indicate that 108,000 engineers—6 percent of the total—had worked for their current employer 12 months or less and had been in the same occupation a year earlier (Table 11). The proportion for engineers was about the same as that for all professional workers (7.0 percent) and all employees (6.2 percent).

Transfer Patterns

Information about occupations into which, and from which, engineers transfer is useful because it has the potential to identify career ladders and sources of supply. Though based on a small sample, some information is available from the January 1987 CPS.

Table 10. Comparison of Data on Entrants (percent of total employment)

Occupation	Movements into Occupations		
	Total	Transfers from Other Occupations	Not Working
Total employed, age 16 & over			
January 1987 CPS	16.3	8.1	8.2
Merged CPS	19.0	8.3	9.6
Professional specialty occupations			
January 1987 CPS	9.2	4.3	4.9
Merged CPS	10.1	4.4	5.5
Engineers			
January 1987 CPS	6.2	3.4	2.8
Merged CPS	5.8	3.6	2.2
Aerospace engineers			
January 1987 CPS	2.0	.0	2.0
Merged CPS	2.9	.0	2.9
Civil engineers			
January 1987 CPS	4.7	.9	3.8
Merged CPS	2.8	1.0	1.8
Electrical & electronic engineers			
January 1987 CPS	5.8	3.1	2.7
Merged CPS	5.5	3.3	2.3
Industrial engineers			
January 1987 CPS	5.9	4.2	1.7
Merged CPS	6.7	4.2	2.6
Mechanical engineers			
January 1987 CPS	8.4	5.1	3.3
Merged CPS	7.3	5.3	1.9
Engineers, n.e.c.			
January 1987 CPS	8.8	5.3	3.6
Merged CPS	7.7	5.6	2.1

SOURCE: 1988 annual average Current Population Survey data.

Table 11. Employees with Current Employer 12 Months or Less (in thousands)

Occupation	Employed, January 1987									
	Number	Percent	With Current Employer 12 Months or Less							
			Number	Percent	Employed, January 1986				Not working	
					Same Occupation		Different Occupation		Number	Percent
					Number	Percent	Number	Percent		
Total employed, age 16 & over	109,008	100.0	21,658	19.9	6,762	6.2	6,268	5.8	8,845	8.2
Professional specialty occupations	14,336	100.0	2,091	14.6	1,000	7.0	396	2.8	703	4.9
Engineers	1,790	100.0	181	10.1	108	6.0	23	1.3	51	2.9
Aerospace	106	100.0	4	3.8	2	1.9	0	.0	2	1.9
Civil	240	100.0	26	10.8	15	6.3	2	.8	9	3.8
Electrical & electronic	530	100.0	53	10.0	36	6.8	3	.6	14	2.6
Industrial	206	100.0	17	8.3	8	3.9	6	2.9	4	1.9
Mechanical	286	100.0	32	11.2	14	4.9	8	2.8	10	3.5
Engineers, n.e.c.	272	100.0	30	11.0	15	5.5	5	1.8	10	3.7

SOURCE: January 1987 Current Population Survey data.

Table 12 shows the occupations in January 1987 of individuals employed as engineers in January 1986 and identifies occupations into which individuals transferred. The largest group who transferred occupations—slightly over one-quarter—moved into managerial occupations, probably because of promotions. At just under one-fifth, the next largest group moved to different engineering specialties but remained within engineering. These results are very similar to those observed in the 1970 Census over the 1965-1970 time period.

Table 13 identifies occupations in which engineers in January 1987 were working a year earlier and identifies occupations from which individuals transferred. The largest groups came from precision production, craft, and repair occupations and other engineering specialities, but there is no clear pattern. However, remember the limitation on the data about individuals transferring from other occupations: students who are working can not be distinguished from qualified engineers.

Net Change Data

Gross movement data describe a complex labor market with large proportions of workers entering and leaving the labor force or moving between occupations. To measure the impact of these movements over time, CPS annual average data were used to identify net changes 1983-1988, by occupation and by age group. For example, the number of engineers age 20-24 in 1983 was compared with the number age 25-29 in 1988. In this case, the number increased and provided a measure of net entrants for that age group. When summed across all age groups registering increases, a measure of net entrants to engineering is obtained. If the change for an age group was negative, however, a measure of net leavers results that, when summed, identifies total net leavers from the occupation. Net change data provide no information about the magnitude of gross movements out of or into occupations, but they do conveniently summarize their impact. Transfers, labor force accessions and separations (including deaths), and migration all contribute to the net result that differs from gross measures but nonetheless is useful for characterizing the behavior of individuals entering and leaving occupations.

Data for engineers show that 53 percent of net entrants to engineering 1983-88 were age 25-29 in 1988 (Table 14). Even though some individuals leave engineering at all ages, the number age 20-24 in 1983 who entered engineering far exceeded the numbers leaving and contributed to the concentration of net entrants within this age group. The result is to be expected since that age group probably includes most individuals who entered

Table 12. January 1987 Occupation of Engineers in January 1986 (in percent)

Major Occupation Group	Employed Engineer in January 1986
Employed in January 1987	100.0
Executive, administrative, & managerial occupations	1.2
Professional specialty occupations	96.7
Non-engineer professional	.3
Engineer	96.4
Same specialty	95.6
Different specialty	.9
Sales occupations	.5
Administrative support occupations, including clerical	.7
Service workers, except private household	.2
Precision Production, craft, & repair occupations	.3
Transportation & material moving occupations	.2

SOURCE: January 1987 Current Population Survey data.

Table 13. January 1986 Occupation of Engineers in January 1987 (in percent)

Occupation	Employed Engineer in January 1987
Employed in January 1987	100.0
Executive, administrative, & managerial occupations	.4
Professional specialty occupations	97.9
Non-engineer professional	.6
Engineer	97.3
Same specialty	96.7
Different specialty	.6
Technicians & related support occupations	.1
Sales occupations	.1
Administrative support occupations, including clerical	.2
Service workers, except private household	.1
Precision Production, craft, & repair occupations	.7
Machine operators, assemblers, & inspectors	.3
Handlers, equipment cleaners, helpers & laborers	.2

SOURCE: January 1987 Current Population Survey data.

Table 14. Net Change Data for Engineers, 1983-1988

	Net Entrants		Net Leavers	
	Engineers	All Workers	Engineers	All Workers
Total employed				
Number (thousands)	359	21,266	157	7,304
Percent				
Total	100.0	100.0	100.0	100.0
1988 Age group				
16-20	1.2	42.8	-	-
21-24	25.9	21.3	-	-
25-29	52.9	14.2	-	-
30-34	6.7	9.0	-	-
35-39	3.1	6.6	-	-
40-44	5.8	4.4	-	-
45-49	4.5	1.6	-	-
50-54	-	-	4.5	.5
55-59	-	-	14.0	9.8
60-64	-	-	36.9	31.8
65-69	-	-	33.8	34.7
70-74	-	-	7.0	12.2
75-99	-	-	3.8	11.0

SOURCE: 1983 and 1988 Current Population Survey annual average data.

engineering immediately after completing college. Net entrants to engineering display a much different pattern from that for all workers. Overall, 43 percent of net entrants were persons age 16-20 in 1988. Among this group, none were in the labor force in 1983 because none were age 16. Thus, in this age group, all are entrants to occupations. Unlike that for net entrants, the pattern for net leavers from engineering is quite similar to that for all workers. Net leavers are greatest in the age group 60-64 followed closely by the 65-69 group. Since most leavers at these ages probably are retiring, the data indicate relatively few engineers retire before age 60.

Are patterns based on 1983-88 net change data for engineers typical of other periods, or are they unique? While differences do exist that reflect differences in labor market conditions for engineers, the 1978-1983 patterns for engineers are remarkably similar to those for 1983-1988 (Table 15): net entrants are concentrated in the age 25-29 group, and few net leavers are observed prior to age 60. Note that total net entrants are greater and less concentrated in 1978-1983 than in 1983-1988. The implication is that as

Table 15. Net Change Data for Engineers, 1983-1988 and 1978-1983

	Engineers Net Entrants 1983-1988	Engineers Net Entrants 1978-1983	Net Leavers 1983-1988	Net Leavers 1978-1983
Total employed				
Number (thousands)	359	468	157	112
Percent				
Total	100.0	100.0	100.0	100.0
1988 Age group				
16-20	1.2	1.0	-	-
21-24	25.9	27.4	-	-
25-29	52.9	42.3	-	-
30-34	6.7	16.7	-	-
35-39	3.1	4.7	-	-
40-44	5.8	4.5	-	-
45-49	4.5	3.4	-	-
50-54	-	-	4.5	.9
55-59	-	-	14.0	17.9
60-64	-	-	36.9	41.1
65-69	-	-	33.8	36.6
70-74	-	-	7.0	2.7
75-99	-	-	3.8	.9

SOURCE: 1978, 1983 and 1988 Current Population Survey annual average data.

more engineers are needed, recent engineering graduates become a smaller proportion of entrants.

Although the patterns of leavers in 1983-1988 and 1978-1983 were similar, net leavers as a proportion of persons age 45 and older increased from 21 percent to 27 percent. Whether the difference reflects the tendency of men to retire earlier or if unique market conditions in 1978-1983 provided incentives to continue working cannot be determined.

Past Versus the Future

A lot of information about engineers during the 1980s has been examined. Growth, gross movements into and out of engineering, movements from and to the labor force—all provide part of the story about engineers that aids in analysis. While recognizing that it is impossible to know positively if data relating to the 1980s are appropriate to a future period, it is a useful exercise to make assumptions about the future and to compare the results with conditions in the past. In this case, demand for additional engineers in 1978-1988 is compared with information for the 1988-1998 period. Demand for additional engineers results from growth and the need to replace workers who leave the occupation. Growth is the easier component to identify.

Based on CPS annual average data, engineer employment 1978-1988 increased by 559,000 while experiencing twice the growth rate for all workers (Table 16). By applying the BLS projected 1988-2000 annual rate of increase to 1988 employment and calculating the change, engineering employment is expected to grow less rapidly but increase by 415,000 in the period 1988-1998. Estimating replacements is more complicated.

Gross flow data indicate about 8 percent of engineers leave the occupation from one year to the next. Some become managers and leave permanently; others retire and leave permanently; yet others leave temporarily to work in another occupation or stop working for a variety of reasons. The group that leaves temporarily not only creates openings, but, when returning to the occupation, constitutes part of the entrants. Because the number of temporary movements out of an occupation are greatly affected by market conditions, they are the most difficult to quantify. It is safe to say, however, that if more engineers are needed, adjustments in the labor market will occur. Perhaps more technicians or college graduates with different specializations will be utilized. Quality may suffer, but adjustments will be made. While recognizing that many opportunities arise because individuals leave engineering temporarily, in this exercise, only opportunities resulting from permanent separations are examined.

To estimate permanent separations, net leavers 1978-1988 were calculated by age groups and summed to yield an estimate of 269,000 engineers who were replaced. To calculate replacement needs 1988-1998, the proportion of net leavers from engineering 1978-1988 in each age group was calculated and applied to 1988 data for the comparable age group. After summing, replacement needs 1988-1998 were determined to increase by 56,000 (20.8 percent), to 325,000. The increase is the result of an aging work force.

In summary, from 1978-1988 there was a need for 828,000 additional engineers. Over the 1988-1998 decade, the need is expected to decline 11 percent to 740,000. One note of caution before concluding that competition for new engineers will ease. In 1978, the population contained 61.1 million persons age 10-24, the source of most additional engineers. In 1988 the population in that same age group had declined to 54.0 million, an 11 percent decline almost identical to the expected 11 percent decline in the need for additional engineers (Table 17).

Table 16. Job Opportunities for Engineers, 1978-1988 and 1988-1998 (in thousands)

	1978-1988	1988-1998	Percent Change
Total	828	740	-10.6
Growth	559	415	-25.8
Replacement needs[a]	269	325	20.8

[a] Estimate of net leavers.

Table 17. Population Age 10-24, 1978 and 1988 (in thousands)

	1978	1988	Percent Change
Total			
Age 10-24	61,101	54,028	-11.6
Age 10-14	18,920	16,627	-12.1
Age 15-19	21,435	18,214	-15.0
Age 20-24	20,748	19,184	-7.5

SOURCE: 1978 and 1988 Current Population Survey annual average data.

Conclusion

Many organizations, including the National Science Foundation and the Center for Education Statistics, are widely known for their efforts to collect and maintain data about engineers. Perhaps less well known is the information available from Bureau of Labor Statistics that has been presented in this paper. Much of the data on occupational transfers and job tenure are unique because they are collected only in special supplements to the Current Population Survey. The last was conducted in January 1987. Unfortunately, another supplement has not been scheduled and, even if the money were found, probably could not be conducted before 1991 because the Bureau of the Census will be concentrating its efforts on the 1990 Census of Population. Now is the time to utilize what is available and to plan for obtaining that which is desired.

APPENDIX A
OCCUPATIONAL EMPLOYMENT STATISTICS SURVEY DEFINITIONS FOR ENGINEERS

Forms used in the Occupational Employment Statistics (OES) survey include these definitions to be used by employers in identifying engineers.

22100 Engineers

Include persons engaged in the practical application of physical laws and principles of engineering for the development and utilization of machines, materials, instruments, processes, and services. Include engineers in research development, production, technical services, and other positions which require knowledge normally obtained through completion of a 4-year engineering college program. Exclude persons trained in engineering but currently working in positions not requiring engineering training.

22102 Aeronautical and Astronautical Engineers

Perform a variety of engineering work in designing, constructing, and testing aircraft, missiles, and spacecraft. May conduct basic and applied research to evaluate adaptability of materials and equipment to aircraft design and manufacture. May recommend improvements in testing equipment and techniques. Exclude sales engineers and report them with the sales workers.

22105 Metallurgists and Metallurgical, Ceramic, and Materials Engineers

Metallurgists and Metallurgical Engineers: Investigate properties of metals and develop methods to produce new alloys, usages, and processes of extracting metals from their ores. Include Physical and Extractive Metallurgists. Ceramic Engineers: Conduct research, design machinery, and develop processing techniques related to the manufacturing of ceramic products. Materials Engineers: Evaluate, plan, and implement processes to develop new materials to meet product specifications, performance standards, and costs. Exclude sales engineers and report them with the sales workers.

22108 Mining Engineers, including Mine Safety

Determine the location and plan the extraction of coal, metallic ores, nonmetallic minerals, and building materials, such as stone and gravel. Work involves: Conducting preliminary surveys of deposits or undeveloped mines and planning their development; examining deposits or mines to determine whether they can be worked at a profit; making geological and topographical surveys; evolving methods of mining best suited to character, type, and size of deposits; and supervising mining operations. Exclude sales engineers and report them with sales workers.

22111 Petroleum Engineers

Devise methods to improve oil and gas well production and determine the need for new or modified tool designs. Oversee drilling and offer technical advice to achieve economical and satisfactory progress. Exclude sales engineers and report them with the sales workers.

22114 Chemical Engineers

Design chemical plant equipment and devise processes for manufacturing chemicals and products, such as gasoline, synthetic rubber, plastics, detergents, cement, paper and pulp, applying principles and technology of chemistry, physics, and engineering. Exclude sales engineers and report them with the sales workers.

22117 Nuclear Engineers

Conduct research on nuclear engineering problems or apply principles and theory of nuclear science to problems concerned with release, control, and utilization of nuclear energy. Exclude sales engineers and report them with the sales workers.

22121 Civil Engineers, including Traffic

Perform a variety of engineering work in planning, designing, and overseeing construction and maintenance of structures and facilities, such as roads, railroads, airports, bridges, harbors, channels, dams, irrigation projects, pipelines, power plants, water and sewage systems, and waste disposal units. Include traffic engineers who specialize in studying vehicular and pedestrian traffic conditions.

22123 Agricultural Engineers

Applying knowledge of engineering technology and biological science to agricultural problems concerned with power and machinery, electrification, structures, soil and water conservation, and processing of agricultural products. Exclude sales engineers and report them with the sales workers.

22126 Electrical and Electronic Engineers

Design, develop, test and supervise the manufacture and installation of electrical and electronic equipment, components or systems, computers and related equipment and systems for commercial, industrial, military or scientific use. Exclude sales engineers and report them with the sales workers.

22128 Industrial Engineers, except Safety

Perform a variety of engineering work in planning and overseeing the utilization of production facilities and personnel in department or other subdivision of industrial establishment. Plan equipment layout, work flow, and accident prevention in department or other subdivision of industrial establishment. Plan and oversee work, study and training programs to promote efficient worker utilization. Develop and oversee quality control, inventory control, and production record systems. Industrial product safety engineers should be included with safety engineers.

22132 Safety Engineers, except Mining

Apply knowledge of industrial processes, mechanics, chemistry, psychology, and industrial health and safety laws to prevent or correct injurious environmental conditions and minimize effects of human traits that create hazards to life and property or reduce worker morale and efficiency. Include industrial product safety engineers.

22135 Mechanical Engineers

Perform a variety of engineering work in the planning and designing of tools, engines, machines, and other mechanically functioning equipment; and oversee installation, operation, maintenance, and repair of such equipment, including centralized heat, gas, water, and steam systems. Exclude sales engineers and report them with the sales workers.

22138 Marine Engineers

Design, develop, and take responsibility for the installation of ship machinery and related equipment, including propulsion machines and power supply systems. Exclude marine architects. Exclude sales engineers and report them with the sales workers.

2199 All Other Engineers

Include all other workers in this category not classified separately above. Please identify in Section V (at the end of this form) all occupations included in this category that are numerically important and require substantial training, or are emerging due to technological changes in your industry.

APPENDIX B
1980 CENSUS OF POPULATION CLASSIFIED INDEX OF INDUSTRIES AND OCCUPATIONS: INFORMATION ABOUT ENGINEERS

The following lists occupational titles for each of the engineering categories. In some cases only an occupational title is listed; in others an industry name and/or industry code is included. When no industry information is provided the occupational title is classified in the category regardless of the industry. If industry information appears, however, the title is classified in the category only if the individual reporting the title also identified the corresponding industry. Additional information and industry names for codes are provided in the 1980 Census of Population Classified Index of Occupations.

Professional Specialty Occupations

044 Aerospace engineers

Aerodynamicist — 352,362
Aircraft designer
Airplane designer
Aviation consultant
Design analyst — 352,362

Designer — 352,362
Dynamicist — 352,362
Engineer
  Aerodynamics
  Aeronautical

  Aerospace
  Aircraft
  Aircraft instrument
  Airplane
  Astronautical

  Aviation — (352)
  Chief, n.s. — 352,362
  Design, n.s. — 352,362
  Field service — 352,362
  Flight — 352,362

  Flight test — 292,352,362,931,932
  Helicopter — (352)
  Propeller — 352,362
  Propulsion — 352,362
  Stress — 352,362

  Supersonic — 352,362
  Test — 352,362
  Test facility — 292,352,362
  Thermodynamics — 352,362
  Transonic — 352,362
  Vibration — 352,362
  Wind tunnel — 352,362
  N.s. — 352,362
Flight analyst — 352,362
Flight dynamicist

Master-lay-out man — 352,362
Physical aerodynamicist — 352,362
Street analyst — 352
Test analyst — 352,362
Thermodynamicist — 352,362

045 Metallurgical and materials engineers

Engineer
  Ceramic — (882)
  Design, n.s. — 270-280
  Foundry process
  Materials — Exc. B

  Metallurgical
  Ore dressing
  Process — 272,280,290,291,300,301
  Refining — 270-280
  Smelting
  Stress — 270,271,280-291,300
  Testing — 270,271,280-291,300
  Welding
Extractive metallurgist

Foundry metallurgist
Metallographer
Metallurgical specialist
Metallurgist

Physical metallurgist

Radiologist — Exc. K,812-840

046 Mining engineers
Engineer
    Design, n.s. — 040,041,050
    Exploration — 040,041,050
    Geological — 040-050
    Geophysical

    Mine development
    Mine exploration
    Mine production
    Mineral
    Mining

    Safety — 040,041,050
Inspector
    Safety, n.s. — 040-050

Mine analyst —040,041,050
Mine expert — 040,041,050
Supervisor
    Safety — 040,041,050
Teachers, exc. elementary and secondary
    Safety — U.S. Bureau of Mines 931

047 Petroleum engineers
Engineer
    Design, n.s. — 042
    Exploration — 042
    Logging — 042
    Mud — 042

    Natural gas — 042
    Oil well — 042
    Petroleum
    Safety — 042
    Test — 042

    Well surveying — 042
    N.s. — 042
Prospecting observer — 042
Safety analyst — 042
Safety director — 042

Seismic observer — 042
Superintendent, oil well services — 042 Supervisor
    Mud-analysis — 042
Well-logging captain — 042

048 Chemical engineers
Blending coordinator — 200
Engineer
    Absorption
    Adsorption
    Chemical — (882)

    Chemical process development
    Chemical test
    Corrosion
    Design, n.s. — 180-201
    Explosives
    Fuels
    Gas combustion
    Lubricating — 200
    Plant — 180-192

    Plastics
    Plating
    Process — 180-200
    Research chemical
    N.s. — 180,181-192,210,211

Manager
    Research — 200
Sand analyst — 270-291,300-370,400,760
Sand technologist — 270-291,300-370,400,760
Supervisor

    Monomer-recovery — 180, 182,191,192
    Poly-area — 180-192
    Polymerization — 180-192
    Technical — 180-192
Technical director — 180-192

049 Nuclear engineers
Engineer
    Atomic process
    Design, nuclear equipment
    Nuclear
    Radiation

    Radiation protection
    Radiological
    Reactor
Radiation officer — FGOV
Radiological-defense officer — FGOV

053 Civil engineers
Engineer
Atomic process
Design, nuclear equipment
Nuclear
Radiation
Radiation protection
Radiological
Reactor
Radiation officer — FGOV
Radiological-defense officer — FGOV

053 Civil engineers
Engineer
Airport — B
Architectural
Asphalt
Base — FGOV

Bridge — B
Building — 441
Building construction
Cadastral
Cartographic

Chief, n.s. — 470
City
Civil
Concrete
Condemnation — 412,900-932

Construction
Contracting — B
County — (B)
Demolition
Design, bridges — (882)

Design, highway
Design, road
Design, n.s. — B
District — B
Drainage — (B)

Erecting — Exc. 310,312,320, 331,332
Flood control — (B)
Forestry
Foundation — (B)
Geodetic

Geological — Exc. 040-050
Highway
Highway research
Highway safety
Hydraulic — Exc. C, 100-392
Hydrographic
Irrigation
Maintenance — 400
Mapping

Materials — B
Municipal — (B)
Process — B
Public health — (840)
Railroad, exc. operating train

Reclamation — (B)
Resident
Road — B
Sanitary
Sanitation, exc. trash or garbage collection

Sewage disposal — (471)
Street — B
Street — B
Structural
Structural steel

Topographical
Traffic — B,400
Transportation
Traveling — B
Water supply
Water treatment plant
Zoning — City Planning Board 901
N.s. — B,H,LGOV 900-932
N.s. — Surveying co. 882
Superintendent, n.s. — 470

054 Agricultural engineers
Engineer
Agricultural
Design, n.s. — 311
Test — 311

055 Electrical and electronic engineers
Assigner, exc. clerical — 441
Assignment man — 441
Circuit designer — 340-350
Communications consultant — 441
Electrolysis investigator — 441

Electronic-parts designer — 340-350
Engineer
  Acoustical
  Audio — 882
  Cable — 441

  Central-office equipment
  Chief, n.s. — 440,441,460
  Circuit design
  Commercial — 440,441,460
  Circuit design
  Commercial — 440,460,462
  Communications

  Computer application
  Corrosion control
  Design, electrical
  Design, electronic equipment
  Design, n.s. — 340-350,440-460

  Dial equipment — 441
  Distribution — 460
  District plant — 441
  Division — 460
  Division plant —441

  Electrical
  Electrolysis
  Electronic
  Electronic systems
  Electrophonic
  Equipment — 441,460
  Facilities — 441
  Guidance and control systems

  Illuminating
  Induction coordination
  Lighting
  Line construction — 460
  Maintenance — 460

  Meter — 460
  Microwave
  Outside plant — 441
  Plant — 441,460,462
  Power generation — 460

  Protection — 460
  Radar
  Radio
  Radio station
  Relay — 460
  Results — 460
  Rural electrification
  Service — 340-350,460
  Signal
  Sound — 800

  Station, n.s. — 440
  Studio operation — 441
  Systems
  Telecommunications
  Telephone — (441)

  Television — (440)
  Testing — Electrical engineering co.882
  Traffic — 441,442
  Traffic circuit — 441
  Traffic routing — 441

  Transmission — 441,460
  Transmitter — 441
  Wire communications
  N.s. — 321,342,382,441,442,460
Engineering analyst

Inspector
  Cable — 460
Line-construction superintendent — 441,460
Radio-interference expert
Supervisor
  Microwave

056 Industrial engineers
Efficiency analyst — (742)
Efficiency expert — (742)
Engineer
  Efficiency
  Establishing methods — C,100-222, 231-392

  Factory lay-out
  Field — H
  Fire prevention — U.S. Army Base 932
  Industrial
  Inspecting — H

  Manufacturing, exc. chief
  Methods
  Production
  Production tool
  Quality control

Safety — Any not listed above
Standards
Time study
Industrial-methods consultant — (882)
Material analyst — 352,362

Material scheduler — 352,362
Medical-safety director — Oxygen 192
Metrologist
Production-control expert
Production-control planner — C,100-392
Production expert

Production planner — Exc. 341
Production scheduler — C, 100-392
Quality-control director
Quality-control expert
Safety coordinator — C,100-392

Safety director — Exc. 040-050,410
Supervisor
Safety — C,100-392
Tool planner — C,100-392
Traffic-rate analyst — C,100-392

Waste-elimination man

057 Mechanical engineers
Engine designer
Engineer
Air conditioning, exc. oper. air cond. systems
Auto research
Automotive — Exc. 351,590, 612-622,750,751

Body — 351
Brake — 351
Combustion
Cryogenics
Design, cooling and heating systems

Design, machine
Design, mechanical
Design, tool
Design, n.s. — 281-291,300-310, 312-332,351,361,370,760

Diesel — Exc. 400
Distribution — 461
Dust control
Equipment — Exc. 441,460
Erecting — 310,312,320,331,332

Field service — 310,312,320, 331,332
Heating, exc. operators of heating systems
Hydraulic — C,100-392
Internal combustion
Mechanical
Mechanical development
Mechanical research
Plant — Mfg. not listed above
Refrigeration, exc. oper. of refrig. systems
Sheet metal — 282

Textile
Tool
Tool and die
Tooling
Utilization — 460-462

Ventilating
N.s. — 292,331,351
Factory expert — 351
Machine designer
Machine-tool designer

Safety analyst — Exc. 040-050
Ventilating expert — (B)

058 Marine engineers and naval architects
Engineer
Boat — (420)
Chief, n.s. — 420
Chief, n.s. — Commercial fishing 031
Hull — 360

Licensed marine — 420
Marine — Exc. fire department 910
Naval
Port — 420
Ship — 420
Tugboat — 420

N.s. — 360,420

Marine architect — (882)

Marine surveyor
Naval architect — (882)

059 Engineers, n.e.c.

Engineer
- Application, exc. computer application
- Biomedical
- Chief, n.s. — 882
- Consulting, n.s. — (882)

- Design, n.s. — Any not listed above
- Distribution — Exc. 460,461
- District plant — Exc. 441
- Environmental
- Factory

- Human factors
- Installation
- Logistics — 292,352,362
- Mathematical
- Medical

- Optical — (372)
- Ordnance
- Packaging
- Photographic
- Pollution-control

- Process — Any not listed above
- Reliability
- Salvage — Exc. 332
- Service — 882
- Staff, field

- Technical, testing — 800
- Technical, n.s. — (882)
- Testing — Any not listed above
- Traffic — Any not listed above
- N.s. — OWN 882 Exc. surveying co.

- N.s. — 730, FGOV 900-932

ADAPTABILITY IN CHEMICAL ENGINEERING

J. S. Watson
Oak Ridge National Laboratory

The adaptability issue for chemical engineering (the ability of chemical engineers to be retrained to work in other fields and for those from other fields to be retrained for work in chemical engineering) will be discussed and illustrated in terms of (1) the interdisciplinary origin of chemical engineering; (2) similarities of chemical engineering curricula with those of other professions; (3) the expected future directions of chemical engineering that are likely to include greater involvement of interdisciplinary skills and, perhaps, greater adaptability; and (4) a few actions that could enhance adaptability. Few quantitative facts are presented in this paper, but a brief section suggests ways to quickly gain limited quantitative information on adaptability.

In considering interdisciplinary programs, it is worthwhile to consider first what a single engineering discipline is. There are different ways to approach and answer this question, but the criterion used in this discussion is the proliferation of undergraduate (and in some cases graduate) departments in the discipline at leading universities. Emphasis is placed upon the undergraduate programs, where the bulk of the classroom training usually occurs. The existence of undergraduate programs constitutes a recognition that a unique combination of courses makes up an important discipline. For a program to remain a separate department, its graduates must be well accepted by industries, government, and graduate schools. This becomes a validation of the discipline. Fields with only graduate programs can qualify as separate disciplines, but their cases may be weakened if the principal body of knowledge studied by the students occurs in the undergraduate program.

Selection of this criteria implies that there is no unique, and perhaps no optimal, way to divide engineering and science into separate fields. The divisions that we have, and that we will have in the future, result from historical developments. As the existing fields expand and more new disciplines are developed from current specialty subfields and interdisciplinary areas, the boundaries of the "disciplines" will appear to be even more arbitrary. Present chemical engineering training constitutes a useful collection of knowledge that serves important needs in our society, but this is not a unique collection of knowledge that could only be packaged this way. In the future there will be new needs, and new information and concepts will have to be covered in the training of chemical engineers. Some materials will be covered in less detail or even eliminated. Subfields or interdisciplinary programs within chemical engineering may eventually become recognized as separate disciplines.

The Interdisciplinary Origin of Chemical Engineering

Chemical engineering can be viewed as either the youngest of the major traditional engineering disciplines or as the oldest of the major new disciplines. It originated as interdisciplinary programs between chemistry and the extant engineering disciplines. Although chemical engineering activities certainly extend well back into the past, they were usually considered interdisciplinary programs until the early part of this century. Consideration of chemical engineering as a separate discipline was demonstrated by the foundation of the first academic program called "chemical engineering" at MIT in 1888 and the foundation of a separate professional society, the American Institute of Chemical Engineers (AIChE), in 1908 (Reynolds, 1983).

Recognition and acceptance of chemical engineering as a separate discipline increased in the following years and has become almost universal in the United States, although there was in time opposition in to formation of a separate discipline (even though the acceptance occurred more quickly in the United States than in some other countries, some major U.S. universities with strong engineering programs have no chemical engineering program). The establishment of a new discipline is almost certain to raise questions and challenges to existing departments and professional societies (disciplines). In most cases, chemical engineering programs grew out of the chemistry and mechanical engineering departments, but, in at least one case, it was once included in the electrical engineering department. The formation of a new discipline is likely to result when the needs of the new discipline are not met by the existing department and society. Formation of a new department (discipline) can be delayed or prevented by opposition from a strong department head from one of the original departments, but it could also be delayed by such exceptional (perhaps excessive) cooperation from the original departments that there would be little apparent need to create a new department.

The curriculum for training chemical engineers and the relationship of chemical engineering to chemistry and the other engineering professions developed slowly but has become relatively uniform, especially at the undergraduate level, due in part to accreditation activities of the AIChE and continuous interactions between chemical engineering departments. The historic evolution of chemical engineering as an interdisciplinary activity has resulted in the location of several important chemical engineering departments in liberal arts colleges, thus associated with chemistry departments, not with the engineering college (the more common location). Graduates from departments affiliated with liberal arts colleges appear to be treated by employers essentially the same as those from departments affiliated with engineering colleges. However, location of the chemical engineering department in the liberal arts college often places it in closer association with the chemistry department and may result in a greater number of required chemistry hours in the curriculum.

The traditional interdisciplinary nature of chemical engineering is also illustrated by its frequent association with other engineering fields in joint departments. (This, of course, also reflects the relatively small size of many chemical engineering departments.) Several chemical engineering departments are, or have been, combined with material science (metallurgical engineering), nuclear engineering, petroleum engineering, or other small programs. As several of these programs became established as separate engineering departments, other newer programs have become associated with chemical engineering departments and are sometimes included in the department titles—for example, chemical and bioprocessing engineering and chemical and polymer engineering. To some extent this reflects only the use of the chemical engineering department to foster new departments that may eventually become separated into new and separate departments. The changing combination of names also reflects the evolution of emphasis in chemical engineering education and practice. The strong emphasis on petroleum and petrochemical processing in chemical engineering may fade somewhat during the coming decades as new areas of "processing" become increasingly important.

The interdisciplinary nature of chemical engineering is also illustrated in its graduate programs. Many chemical engineering departments now include important faculty members who were educated in other disciplines as well as in chemical engineering. There are even notable examples of faculty trained solely in related fields rather than directly in chemical engineering. There have been significant transfers to the profession from related fields such as chemistry. Transfers enter graduate studies in chemical engineering after obtaining B.S. or higher degrees in a related field. The motivations for such transfers include discovery of an interest in chemical engineering (lack of public understanding of what chemical engineers do and study has always been a problem), better employment prospects, or higher salaries. The number of transfers has raised some concern for erosion of quality in "traditional" chemical engineering training, but the transfers also indicate the profession's commitment to interdisciplinary work and the relation of chemical engineering to other fields. The interdisciplinary aspects may even be more historical or "traditional" to chemical engineering than any particular courses that have become known as "traditional chemical engineering" during the last four or five decades. The relatively small number of chemical engineers makes it necessary for them to interact more with other engineering disciplines than for engineers from the more populated disciplines. The relation of chemical engineering to chemistry remains important and usually results in interactions in joint programs, perhaps more in industrial situations than in the university departments.

Curriculum for Chemical Engineering

Undergraduate chemical engineering students obviously study far more chemistry than any of the other major engineering disciplines, but the amount varies from school to school. More than five semesters of chemistry are required by essentially all schools; some require considerably more. The curricula require additional technical electives, which can include additional chemistry courses. Chemical engineering students do not take as much chemistry as chemistry majors, but the difference need not be great if the student chooses a school with a high chemistry curriculum and takes chemistry for some of the technical electives.

Since engineering students take most courses in their undergraduate major only during their junior and senior years, there is much similarity between curricula for all fields of engineering. Furthermore, many chemical engineering courses are similar to related courses in mechanical and civil engineering: the undergraduate curriculum usually includes one semester of thermodynamics and at least one semester of fluid mechanics and heat transfer. Graduate students may take courses in such topics from other departments, and some graduate courses in these topics may be offered for joint (either department) credit. Courses in the chemical engineering department are more likely to have a chemical engineering emphasis and highlight "process" applications, but much of the material may be similar to that taught to mechanical and civil engineering students. Chemical engineers and other engineering graduates probably can be highly adaptable for work in those topics covered by both departments. With normal in-house training, combustion, hydrodynamics, and heat transfer phenomena probably could be handled by either chemical or mechanical engineering graduates. However, chemical engineering graduates would require extensive retraining to handle problems in mechanical engineering topics such as engine design and structural analysis. Similarly, while chemical engineers could quickly be trained to work on many topics in environmental engineering, extensive retraining would be required for structural and construction positions in civil engineering.

Graduate education tends to become more highly specialized, and adaptability may be less evident since graduates may not be just chemical engineers, chemists, or other types of engineers, but true specialists in narrow subfields. The differences between graduate programs in chemistry, chemical engineering, and other engineering departments may in some cases become even less clear. Several specialty areas for chemical engineering research and graduate study—such as specific problems in chemical and physical design of equipment or even integrated processing facilities—are usually considered unique to chemical engineering. Other research areas strongly overlap traditional chemistry studies, especially areas of physical chemistry, or other engineering disciplines: thermodynamics and phase equilibria studies are active in both chemistry and chemical engineering programs; surface and interfacial studies may be found in chemical engineering, chemistry, physics, and even other engineering departments. Several areas of materials production and materials properties are actively studied in chemistry and material science as well as chemical engineering. The recent increase in studies related to pollution control and waste management have resulted in similar or related studies in environmental (sometimes civil) engineering.

The recent growth of interdisciplinary university "centers," sponsored by the National Science Foundation (NSF) or other organizations, increases the likelihood of graduate students from different departments working together on different aspects of a problem. It is likely that the graduate students from these programs will be selected by future employers first for their special knowledges; the name of the department granting their degree may be a secondary consideration. There are examples (but no known statistics) of major professors who hold degrees in other fields but teach and supervise research within chemical engineering departments. There are probably similar examples of professors with chemical engineering degrees working in other departments. The mixing of such degrees is likely to be even more extensive in industry, which is concerned with the particular department from which the degree was obtained.

The importance of industrial in-house training and experience on adaptability of engineers and other professionals should be noted. Many engineers work in activities that call for technical understanding but do not necessarily require extensive, specific initial knowledge of the industry: intensive specialized knowledge may be required but so specialized that firms prefer to use their own in-house training to supplement the more general training from the universities. In such industries so much specific in-house training could lead to considerable interchangeability of new graduates. One indication of the use of people from other disciplines in chemical engineering is the number of members of the AIChE who hold degrees in other fields (AIChE, 1987): in 1986, these represented 7.6 percent of the members with B.S. degrees and 16.9 percent of the members with M.S., and Ph.D., degrees. Membership in the AIChE is one indication that a person feels that he/she is are engaged in chemical engineering activities. Engineers working in a particular industry for many years acquire on-the-job knowledge that eventually can become more important than their initial training. Adaptability usually declines with time or age if the engineer becomes more specialized, but the importance of the initial education also becomes less important than the knowledge acquired after formal schooling.

Expected Trends in Chemical Engineering

A recent study by the National Research Council (1988) has attempted to predict future trends in chemical engineering and identify areas where new growth is most likely to occur. The study, which followed a similar study of chemistry and occurred during the period of unusually high unemployment and concern among chemical engineers, can be viewed as an assessment of opportunities for chemical engineers as well as a prediction of new growth areas—namely, biotechnology and biomedicine; electronic, photonic, and recording materials and devices; and microstructured materials. An interesting aspect of these areas is that they all involve interdisciplinary activities. Biotechnology clearly involves biologists (perhaps several types) and biochemists. Chemical engineers are expected to play major roles in bringing the fruits of biotechnology to process scale facilities, but the roles of other disciplines are not likely to be diminished by participation of chemical engineers. If chemical engineering participation means that more products of biotechnology reach the marketplace earlier and at lower prices, the roles of all participants in biotechnology will be enhanced.

There is some study of biomedical engineering (sometimes under different names) in older traditional engineering disciplines. In fact, journals and conferences organized by several engineering societies show signs of competition as each society stakes out its role in a "hot" field, but the competition is more apparent than real. Each traditional engineering discipline has a different view and interest in the field; so the overlap of interests is limited. Electrical engineers may think of biomedical engineering in terms of instrumentation and sensors for monitoring medical phenomena and detecting and evaluating diseases. Mechanical engineers and material scientists may think of biomedical engineering in terms of artificial hearts, artificial limbs, and other devices that would correct medical problems. Others may think principally of new "engineered" drugs, biochemical sensors, analyses of body fluids, and chemical treatments—areas where chemical engineering is more likely to play major roles in biomedicine. Thus the different interests of various engineering disciplines are more important than the overlap.

Electronic, photonic, and recording materials currently are enjoying rapid growth that is expected to continue into the foreseeable future. Expected new developments will permit greater use of newer optical materials (systems) for information storage and faster electronic materials to replace the traditional silicon-based materials and accelerate the growth in the use of these materials. The combination of knowledge in chemistry and processing technologies included in the training of chemical engineers is important in this field and will become more important as the demand increases for higher purity materials and higher productivity processes. However, the role of electrical engineering, materials science, chemistry, and so forth will remain very important and will not decline. Participation by chemical engineers should enhance the productivity and quality of these products, increase the importance of the material, and thus enhance the participation of all disciplines involved.

Microstructured materials—molecular and physical structures that are designed and constructed for specific high performance—include structured polymers with selected molecular weight ranges and molecular orientation to maximize or optimize one or more desired properties such as strength or toughness. The high expected growth rate in microstructured materials results partially from the growing demand for new materials in new high technology products and partially from a growing awareness that demand for high quality products (even traditional products) is more likely to show higher growth than for lower quality products.

Adaptability in Periods of Economic Stress

Adaptability of different engineering disciplines becomes most important in periods when there are surpluses or shortages in a particular discipline. One way to assess the adaptability of chemical engineers and the potential for graduates from other fields to enter chemical engineering would be to compare the effects of economic problems (high unemployment and even layoffs) on chemical engineers and other graduates. If the graduates are easily exchangeable, unemployment in one field should affect all interchangeable graduates. If there is no interchangeability, high unemployment should affect the disciplines differently. There are, however, drawbacks that limit the information that can be gained via this approach.

The high demand (relative to supply) for chemical engineers for several decades prior to the early 1980s allowed little change in the employment rates; the percent of chemical engineering graduates not receiving job offers was too low for effective analysis. The unemployment rate in 1980 was 0.4 percent (American Institute of Chemical Engineers, 1987). This situation changed as enrollments increased sharply during the late 1970s and the demand dropped during the early 1980s (Table 1). This one major perturbation in unemployment of chemical engineers was disturbing to the profession, but it provides only limited information on how employment problems in chemical engineering and other fields are correlated. The major perturbation occurred when a number of factors affected other professions as well as chemical engineering, and changes in their employment picture do not necessarily indicate that chemical engineers were taking positions in other fields.

The one correlation that probably does have important meaning lies in the relations between employment in chemical and petroleum engineering, especially closely related fields in which interchangeability is very high. The major decline in employment of chemical engineers in the early 1980s resulted principally from a sharp decline in the energy industries. The two notable areas were the private petroleum industry and the partially federally sponsored (but largely privately operated) synfuel programs. The petroleum industry has long been a major employer for chemical engineering graduates, and the synfuel program created a major perturbation in the demand for chemical engineers. The profession is small enough that a single major new demand such as this can have important effects upon supplies. To meet this demand, as shown in Table 2, chemical engineering departments significantly increased enrollments and graduates from them (Rawls, 1989). In spite of efforts to maintain quality questions remain as to whether university departments

Table 1. Unemployment Rate Among Chemists and Chemical Engineers (in percent)

	1980	1981	1982	1983	1984	1985	1986
Engineers[a]	0.4	0.5	0.9	4.3	3.2	2.8	3.2
Chemists[b]	0.9	1.1	1.5	2.2	1.7	1.4	1.7

[a]Data from American Institute of Chemical Engineers, 1987. *Economic Survey Report*. New York: AIChE.
[b]Data from Rawls, Rebecca L. 1989. Facts & figures for chemical R&D.*Chemical and Engineering News*, 67(34):56.

that expanded their enrollment did so without loss in quality. This expansion in undergraduate enrollment was not reflected strongly in the graduate programs, probably because high demand and salaries attracted B.S. graduates into industry and reduced the normal fraction continuing their studies in graduate schools.

The subsequent employment decline in the petroleum industry extended quickly into the related petrochemical industry. Many of the largest petrochemical facilities are owned by petroleum companies, and financial needs alone were sufficient to couple the two industries. Not only were new graduates not receiving sufficient offers for employment, but experienced chemical engineers were being terminated or forced (often enticed) into early retirement. Comparisons between the unemployment rates for chemical and petroleum engineers during this period are similar. The unemployment rate for chemists also peaked in 1983, but the peak was not as great (Table 1). Other engineering disciplines associated more closely with defense or construction programs have experienced more periods of serious declines, but the chemical engineering community viewed this period very seriously both because of the number of pcople involved and because of the fear that it could signal a major change in the chemical engineering market pattern. The perturbations in other major engineering disciplines during this period were not as serious, at least relative to previous periods, and apparently caused less trauma.

One long-term effect of this perturbation in employment prospects for chemical engineers has been a significant decline in undergraduate enrollment in chemical engineering (as well as petroleum and chemistry) programs (Rawls, 1989). To a limited degree this has probably been beneficial since the enrollment may have grown too rapidly during the previous years. Recent enrollment figures indicate that the situation has stabilized, with total enrollment at levels similar to those of a decade ago. There is now reason to be concerned that enrollments may be inadequate for growing needs in new and important areas of technology where chemical engineers are expected to be needed.

The most serious effects of the employment perturbation were on the individuals directly involved, experienced engineers who were laid off or retired early and new graduates who did not find jobs immediately after graduation. There is little information on the overall effects of this period and the subsequent fate of the individuals. Personal contacts reveal retired individuals who have started small consulting businesses (some were successful and some were probably not successful), who found jobs with other companies, who remained retired, and who went into other businesses. Even without statistics on the fate of these engineers, it is probably fair to say that this reservoir of engineering talent and experience is not utilized as completely after the early retirements as before.

The fate of young new graduates is equally difficult to assess. In some cases, failure to find suitable employment sent the better students to graduate schools. Enrollment in graduate programs in chemical engineering had declined during the period of high demand for B.S. graduates, but rebounded after the demand declined (Table 2). That could be a good result; but some of the best graduates of that period may have been discouraged from entering graduate schools because immediate job offers were understandably more enticing than the risk of not finding a job after graduate school. For other less fortunate students, there were neither jobs in chemical engineering nor suitable positions in graduate school. Many, perhaps even most, of those students may be lost to the profession: some went to graduate school in fields where the job market was perceived to be more favorable; others simply accepted positions in other fields that probably did not meet their career goals. Being older than new graduates with no more experience and a record of underemployment or even unemployment, these individuals are not likely to be looked upon with favor by employers.

The situation of the more specialized petroleum engineers probably has been similar. Since the petroleum industry was one epicenter of the downturn, unemployment among oil companies alone was particularly serious and enrollment in petroleum engineering plunged even more drastically than in chemical engineering and compensated somewhat for the reduced demand. The long-term outlook for the petroleum and petrochemical industries in this country may be constrained by our limited supply of crude oil. A shift of secondary processing facilities toward the oil-producing countries has started, and it is unclear how many of these activities can be maintained in the United States. Petroleum engineering is an interesting example of engineering specialization that has flourished but faces problems because of its relatively narrow focus.

Table 2. Degrees Awarded in Chemical Engineering and Chemistry, 1968-1987

Year	Chemistry			Chemical Engineering		
	B.S.	M.S.	Ph.D.	B.S.	M.S.	Ph.D.
1968	10,847	2014	1757	3211	1156	367
1969	11,807	2070	1941	3557	1136	409
1970	11,617	2146	2208	3720	1045	438
1971	11,183	2284	2160	3615	1100	406
1972	10,721	2259	1971	3663	1154	394
1973	10,226	2230	1882	3636	1051	397
1974	10,525	2138	1828	3454	1045	400
1975	10,649	2006	1824	3142	990	346
1976	11,107	1796	1623	3203	1031	308
1977	11,322	1775	1571	3581	1086	291
1978	11,474	1892	1525	4615	1237	259
1979	11,643	1765	1518	5655	1149	304
1980	11,446	1733	1551	6383	1271	284
1981	11,347	1654	1622	6527	1267	300
1982	11,062	1751	1722	6740	1285	311
1983	10,746	1604	1746	7145	1304	319
1984	10,704	1667	1744	7475	1514	330
1985	10,482	1719	1789	7146	1544	418
1986	10,116	1754	1908	5877	1361	446
1987	9,661	1738	1976	4983	1184	497

SOURCE: Anonymous. 1986. Unemployment slightly higher for chemists in the past year. *Chemical and Engineering News*, 64(26):23-27.

Increasing the Adaptability of Chemical and other Engineers

The key to increasing the adaptability of chemical engineers for work in other fields is to include broader and additional training in the chemical engineering curriculum. The knowledge taught in the colleges and universities should be as general as is practical, with specific knowledge for particular companies or even some small industries given to the engineers "on the job." However, the expanding technical knowledge appears to have trended toward specialization, not generalization.

Specialization of both scientists and engineers has resulted from simple recognition that so much specialized knowledge is required for professional practice that a general training curriculum in science or engineering would be impractical. Continuing growth in human knowledge is likely to increase the need for specialization. Training for research in the sciences has become sufficiently specialized that department names alone do not cover or indicate the numerous special subfields: for instance, a chemist may be considered an analytical chemist, an instrumental analysis chemist or a spectroscopist.

The engineering professions have resisted this degree of specialization, perhaps because more professional engineering practices involve activities other than research and use B.S., and M.S., as well as Ph.D. degrees. Nonetheless, the pressure has been high for "cramming" more knowledge into the engineering curriculum. The chemical engineering curriculum includes heavy loads of classwork and laboratories, and the need for adding more material becomes evident almost every year. In biology or materials sciences, there appears to be little room for additional material needed for projected new growth areas. There have been calls for a 5-year curriculum or adoption of the M.S. or a higher degree as the lowest professional degree in chemical (or other areas of) engineering to allow more information in the curriculum, but those calls have been resisted.

As long as increases in the curriculum are resisted, it is not likely that significant quantities of new core material can be added to meet the projected needs of the profession and increase the adaptability of chemical engineers. The ultimate way to include additional information into a curriculum is to develop better ways to systemize the information. This requires better insight and understanding that identify similarities and allow several phenomena to be studied together. Systemization can involve development of theories and models that allow complex behavior of systems to be described in new and more concise forms. Such advances must always be sought, but they require ingenuity. The rate of such advancements cannot be predicted.

Recent developments in computer and information technologies suggest new ways that more information can be taught in a given period of time. The full potential of rapidly improving computer capabilities is not known, but obviously could impact the teaching of chemical (and other areas of) engineering: numerical solutions and other descriptions of chemical engineering phenomena may replace verbal, time-consuming laboratory demonstrations. The results of student access to large data bases or artificial intelligence systems is more difficult to assess. Our ability to incorporate these new computer and information technologies into the training process has lagged developments in the hardware. Developing effective software for improving and streamlining scientific and technical education is especially difficult in fields, like chemical engineering, where the number of potential students is not large enough to justify large efforts in software development.

Another obvious case of a non-optimal factor in a chemical engineering curriculum is uncontrolled duplication. Some duplication is necessary to stress selected subjects, but duplication should occur only for specific and intended reasons. Generally, duplication in course material taught by individual departments has been eliminated. Exchange of information between chemical engineering departments throughout the country allows departments at each university to see how other universities are handling these problems and to assess the value of different approaches to their particular department with its unique resources (staff and facilities), goals, and philosophy.

Perhaps the major source of uncontrolled duplication of instructional material in the chemical engineering curriculum occurs because of incomplete cooperation between different departments within the universities, an important consideration for any new engineering discipline that maintains significant portions of its classwork in other departments. Chemistry can be taught to chemical engineers either as part of the regular offering of the chemistry departments or in special classes for chemical engineers (taught either by faculty from the chemistry department or by chemical engineering faculty). By having the chemical engineering students participate in chemistry classes within the chemistry department's standard curriculum for chemistry majors, the instruction is more likely to be carried out by those best informed in the subject and the competition with chemistry majors helps ensure that high standards are maintained.

However, the packaging and organization of chemistry cannot be optimized for presentation to both chemistry and chemical engineering majors. This leads to duplication when students study similar material both in the chemical engineering thermodynamics course and in physical chemistry. Both the lesser number of chemical engineering students and the appropriate primary focus of the chemistry curriculum upon the needs of chemistry majors means that the chemical engineering curriculum must be arranged around the chemistry curriculum rather than optimized for chemical engineers: the content of chemistry courses cannot be limited only to the subjects of most importance to chemical engineers. When chemical engineering students take one fewer semester of a sequence (such as organic chemistry) than chemistry majors, they must take the material as it is packaged by the chemistry department.

The obvious solution to the problem is to maximize cooperation between departments, but this does not mean that the chemistry departments will organize their programs to best suit the needs of chemical engineering students. Their responsibility remains first with their own, more numerous, chemistry majors; cooperation only means that they consider the needs of the chemical engineers and work to find suitable compromises. If a department becomes involved in cooperation with several other departments, the needs of several programs may become compromises.

Suggestions for Improving Data on Adaptability

There are few ways to assess how chemical engineers and other disciplines are interchanged in industry or government. That would require precise definitions of what is, and is not, chemical engineering and leave areas of uncertainties and disagreement. Furthermore, it would be necessary to have information from industry that is unavailable and would be difficult to obtain. Instead, to improve our understanding of the current exchangeability, a more modest approach is recommended. Though less direct and less complete, it has the potential for providing partial answers quickly and with modest effort. Questionnaires can be sent to the major chemical engineering departments asking for the following information:

1. How many faculty members do you have, and how many hold one or more degrees in a field other than chemical engineering? What degrees and in what other fields?
2. How many faculty members in your department hold no degree in chemical engineering (or hold their highest degree in another field)?
3. How many graduate students do you have? How many are working on research under an advisor or co-advisor from another department? How many of your faculty are advising students from other departments?
4. How many of your graduate students hold undergraduate degrees in fields other than chemical engineering?
5. How many of your seniors during the last (few) years entered graduate schools in fields other than chemical engineering?

A simple survey such as this will not answer the larger questions concerning adaptability of chemical engineers, but it can assess how chemical engineers are exchanged with other disciplines at the universities. This represents a modest effort that would provide a basis for estimating how much campus interaction is laying a basis for later interchange of chemical engineers. There should be some similarities in the exchange of engineers in industry.

A limited insight into the use of nonchemical engineers can be obtained by assuming that those working as chemical engineers are likely to become associated with the AIChE and asking that society how many members/associate members have degrees in other fields or no degree in chemical engineering. It would be more difficult to estimate how many chemical engineers are working in other fields, but similar checks could be made of memberships in other societies.

References

American Institute of Chemical Engineers. 1987. *Economic Survey Report.* New York: AIChE.

Anonymous. 1986. Unemployment slightly higher for chemists in the past year. *Chemical and Engineering News* 64(26):23-27.

National Research Council, Committee on Chemical Engineering Frontiers. 1988. *Frontiers in Chemical Engineering: Research Needs and Opportunities.* Washington, D.C.: National Academy Press.

Rawls, Rebecca L. 1989. Facts & figures for chemical R&D. *Chemical and Engineering News* 67(34):56.

Reynolds, T. S. 1983. *75 Years of Progress—A History of the American Institute of Chemical Engineers, 1908-1983.* New York: American Institute of Chemical Engineers.

NUCLEAR ENGINEERING CASE STUDY

David M. Woodall
Idaho National Engineering Laboratory

Abstract

From its infancy in the '50s, the discipline of nuclear engineering has grown into a mature field with applications as diverse as commercial nuclear power, radiation interaction with matter, fusion energy research, and radiological applications to biology. Early nuclear engineering activities were performed principally by physicists, chemists, and mechanical or chemical engineers. The current nuclear engineering education process produces engineers who are broadly prepared in fundamental engineering sciences, yet uniquely able to solve problems involving nuclear processes and nuclear forces. While many nuclear engineers are employed in the commercial nuclear power industry, the discipline of nuclear engineering produces engineers with a breadth far beyond that application. The paper is broken down into sections on the emergence, development, and current status of nuclear engineering as a field with a final section on the future of the discipline.

Introduction

Any study of nuclear engineering would be incomplete without a review of the history and status of commercial nuclear power, which has been the primary driving influence on the field. One should also note that both the federal regulation of the nuclear power industry and the public perception of the risk and benefits of nuclear electric power have had a significant influence on the commercialization of the technology. Comments on the various impacts of such factors on the nuclear engineering discipline are also included here.

In many ways, nuclear engineering is an ideal test bed for studying the flexibility of engineers, for it is a discipline which was born only about 30 years ago, and the industry that it supports has seen a number of major upheavals in the intervening period. Engineers and scientists from many disciplines have entered the nuclear engineering field in the past three decades and nuclear engineers have seen the thrust of their work take a number of radical changes.

Emergence of Nuclear Engineering as a Field

The field of nuclear engineering had its roots in the development of commercial nuclear power (Dawson, 1976). The U.S. Atomic Energy Commission (AEC) was formed in 1946 by an act of Congress, with the mission to develop electric power from nuclear energy. The infrastructure of the nuclear weapons development complex from World War II was used for the construction and testing of a number of reactor types. In 1948, the AEC approved a plan to build and test four reactor projects, including the Materials Test Reactor (MTR) and the Experimental Breeder Reactor (EBR-1) to be built at the National Reactor Testing Station (presently the Idaho National Engineering Laboratory). Both the Argonne and Oak Ridge Laboratories also were actively involved in the nuclear reactor design and development process.

The continued development of commercial nuclear power was enhanced by the passage of the Atomic Energy Act of 1954, which established the early development of commercial nuclear power as a national objective, encouraged enhanced industrial participation in the process, and led to the birth of the commercial nuclear industry. In parallel with this process, the U.S. Navy engaged in a development program for nuclear-powered propulsion. The commercial nuclear power and nuclear propulsion programs were synergistic, with a cross-fertilization of ideas and technologies.

Most early work in nuclear engineering was done at the national laboratories under the control of the U.S. government. Research workers in the field had backgrounds from physics, chemistry, or engineering, typically chemical or mechanical. The declassification of this technology in the mid '50s and the potential for application of nuclear processes to energy production attracted many academics to enter the field as researchers.

A number of academic programs for the study of nuclear engineering were formed initially within existing engineering departments. Between 1955 and 1960 approximately 50 universities had established nuclear engineering degree programs, initially at the graduate level, with an emphasis on neutron diffusion and transport and reactor physics. Table 1 outlines a typical curriculum for the M.S. in nuclear engineering (NE) from the mid

Table 1. Typical M.S. Nuclear Engineering Curriculum in the 1960s

3 Hrs	Nuclear and Radiation Physics
6 Hrs	Reactor Physics (neutron diffusion and kinetics)
3 Hrs	Laboratory: Radiation Interaction with Matter
3 Hrs	Laboratory: Reactor Phenomenology
3 Hrs	Reactor Technology
3 Hrs	Nuclear Power Thermal Hydraulics
3 Hrs	Radiation Shielding
6 Hrs	Advanced Engineering Mathematics (PDE's, integral functions)

'60s. The emphasis on the phenomenology of neutron interaction with matter and reactor criticality was driven by the physics background of many faculty in the field. The curriculum of that period was principally based on physics, rather than engineering. While early nuclear engineering faculty members had the variety of backgrounds previously noted for the early workers in the field, physics backgrounds were in the majority. Many early nuclear engineers obtained graduate training in the discipline without strong engineering sciences preparation at the undergraduate level. Missing from such backgrounds would be one or more of the fundamental engineering sciences—heat transfer, engineering thermodynamics, fluid flow, and engineering materials—as well as a significant design experience.

Much of the early work of nuclear engineers—including nuclear fuel cycle studies and nuclear fuel management, requiring neutron diffusion and nuclear criticality calculations—was in the areas of reactor nuclear design. Fuel performance, radiation health physics, and nuclear power plant operations were emphasized, areas well suited to the nuclear engineers with a physics background. Because the nuclear engineering discipline has only recently emerged, it is true even today that the senior faculty members of nuclear engineering departments often have degrees from fields other than nuclear engineering. However, retirements during the next 10 years will cause a shift to a preponderance of nuclear engineering-trained faculty.

Development of Nuclear Engineering

The growth of commercial nuclear power in the '60s and '70s led to a strong demand for trained nuclear engineers. The nuclear steam supply system (NSSS) vendors required nuclear engineers for nuclear core design, nuclear thermal hydraulic design, nuclear materials evaluation, and nuclear safety. Growing concern within the federal establishment about the availability of trained nuclear engineers to develop and manage commercial nuclear power led the AEC to take two actions: (1) a manpower assessment program for nuclear engineering through Oak Ridge Associated Universities (ORAU), which continues to this day,[1] and (2) establishment through ORAU of a nuclear science and engineering fellowship program in order to attract high-quality scientists and engineers into the field.

During this period the electric utilities embraced commercial nuclear power as the electric power source of the future. Utilities rushed to include nuclear power in their planning, and the U.S. commitment by 1975 was for 225 nuclear power plants [the remainder of the world had a somewhat larger commitment because, while the United States is perceived as having adequate alternative energy resources, including major coal deposits, much of the remainder of the free world has no such option. Thus their commitment to nuclear power reduced their reliance on imported energy, specifically petroleum. The International Atomic Energy Agency (IAEA), the lead agency in promoting the international use of nuclear energy, has been active since its inception in the development of standards for nuclear power operations and the education and training of nuclear professionals[2]].

The electric utilities that ordered nuclear power plants established a hiring program for nuclear engineers in order to manage this new technology. Their primary concern was in radiological control, nuclear regulatory requirements, and nuclear fuel management. It is noteworthy that the best success records have been achieved by those utilities that established senior management commitment and involvement with the construction and operation of their nuclear plants; senior utility executives who accepted nuclear power

---

[1] An annual report on the status of nuclear engineering is published under the auspices of the U.S. Department of Energy and the U.S. Nuclear Regulatory Commission, which inherited the AEC role for nuclear power. The most recent version is *Nuclear Engineering, Enrollments and Degrees, 1987* (DOE/ER—0370), Washington, D.C.: Government Printing Office, 1988.

[2] See, for example, *Radiation Engineering in the Academic Curriculum*, Vienna, Austria: IAEA publishers, 1975.

plants as part of their electric power mix without direct management involvement fared less well. This growing demand for nuclear engineers within the electric utilities caused additional pressure on nuclear engineering departments for production of graduates. By the mid '70s, most universities with nuclear engineering departments had initiated undergraduate degree programs, making the B.S. degree in nuclear engineering more commonplace in the industry. As noted previously, the nuclear engineers of the '60s and early '70s were typically trained at the undergraduate level in another discipline, followed by a 30-hour M.S. program in nuclear engineering. With the advent of the B.S. in nuclear engineering, it was possible to take the core curriculum in mechanical or chemical engineering and modify the junior and senior years to include a specific emphasis in nuclear.

It should be noted that the B.S. in nuclear engineering covers the same basic engineering science as a B.S. in mechanical engineering (ME). The upper-level course work for a B.S. in nuclear engineering is outlined in Table 2. The following topics are usually included in the freshman and sophomore years of both curricula: physics through modern physics, chemistry, mathematics through calculus and differential equations, thermodynamics, fluids, heat transfer, statics, electronics, materials science, and engineering economics. In addition, the nuclear engineering student studies advanced atomic and nuclear physics, radiation interaction with matter, radiation measurements and instrumentation, statistics of counting, neutron and reactor physics, nuclear thermal hydraulics, radiation shielding and safety, as well as nuclear licensing and regulation. There is an emphasis in both the ME and the NE curriculum on laboratory work and engineering design, the synthesis of engineering sciences for a solution to an open-ended problem.

While the B.S. mechanical engineer and the B.S. nuclear engineer have a common heritage, there are many tasks in the nuclear power industry for which the B.S. nuclear engineer is uniquely qualified. Areas that require the nuclear engineering degree include radiation shielding and health physics, reactor neutronics, reactor instrumentation and control, and radiological waste management. Common areas of work include project management, operations, and engineering design of components and systems.

Recognizing the strong demand for nuclear engineering in the early '70s, a number of mechanical and chemical engineering departments initiated a nuclear engineering option program for their undergraduate students. These programs produce a B.S. in the relevant engineering discipline, with a minor in nuclear engineering. Individuals with such degrees

Table 2. Modern Upper-Level B.S. Nuclear Engineering Curriculum

Semester I	Semester II
Junior Year:	
Engineering Math	Statistics
Atomic and Nuclear Physics	Reactor Physics I
Engineering Economics	Fluids and Heat Flow Lab
Reactor Technology	Radiation Measurements Lab
Engineering Materials	Nuclear Materials
Senior Year:	
Reactor Physics II	Nuclear Design II
Nuclear Thermal Hydraulics	Reactor Laboratory
Reactor Safety and Licensing	Fusion Technology
Technical Elective	Technical Elective
Nuclear Design I	

typically use optional technical electives in the undergraduate curriculum to study nuclear technology. Such an engineer has approximately 10-12 credit hours of study in nuclear technology and is well prepared to work as a general engineer (mechanical or chemical) in project management or operations for the nuclear utility industry. However, he or she is poorly prepared to take on the specific tasks noted in the preceding paragraph of a nuclear engineer. The industry experience of using the B.S. engineer with only a nuclear option for nuclear engineering tasks has met with little success.

Present Status of Nuclear Engineering

Undergraduate enrollment in nuclear engineering peaked in 1977, just prior to the Three Mile Island (TMI) nuclear accident, which led to a major reevaluation of the nuclear power industry in this country, as well as a reexamination of the curricular content of the typical nuclear engineering degree. While the degree of individual injury that resulted from the TMI accident was minor compared to routinely accepted hazards in modern life, the public perception of nuclear power was very negative. The regulatory and licensing

The Nuclear Regulatory Commission established a number of safety modifications, or "back-fits," designed to lessen the likelihood of a repetition of the TMI accident. Such modifications made operating plants more complex, increasing the number of required nuclear engineers for operation. A specific position, shift technical advisor, was created for the operating staff that essentially required an engineering degree.

As shown in Figure 1, during this period of increasing demand in operation and licensing for nuclear engineers, the negative public perception of nuclear power led to a steady decline in the enrollment in undergraduate nuclear engineering programs beginning in 1977 (from a high of 2,121 to about 1,300 today) undergraduate enrollment and degree production in nuclear engineering from the early '70s until today. Enrollments began a decline, . Annual degree productivity dropped steadily from a high of 863 in 1977 to about 500 today. Contributing to this steady decline in enrollment was the tendency for engineering colleges with small enrollments in nuclear engineering to drop that program or combine it with larger engineering departments in the late '70s and early '80s.

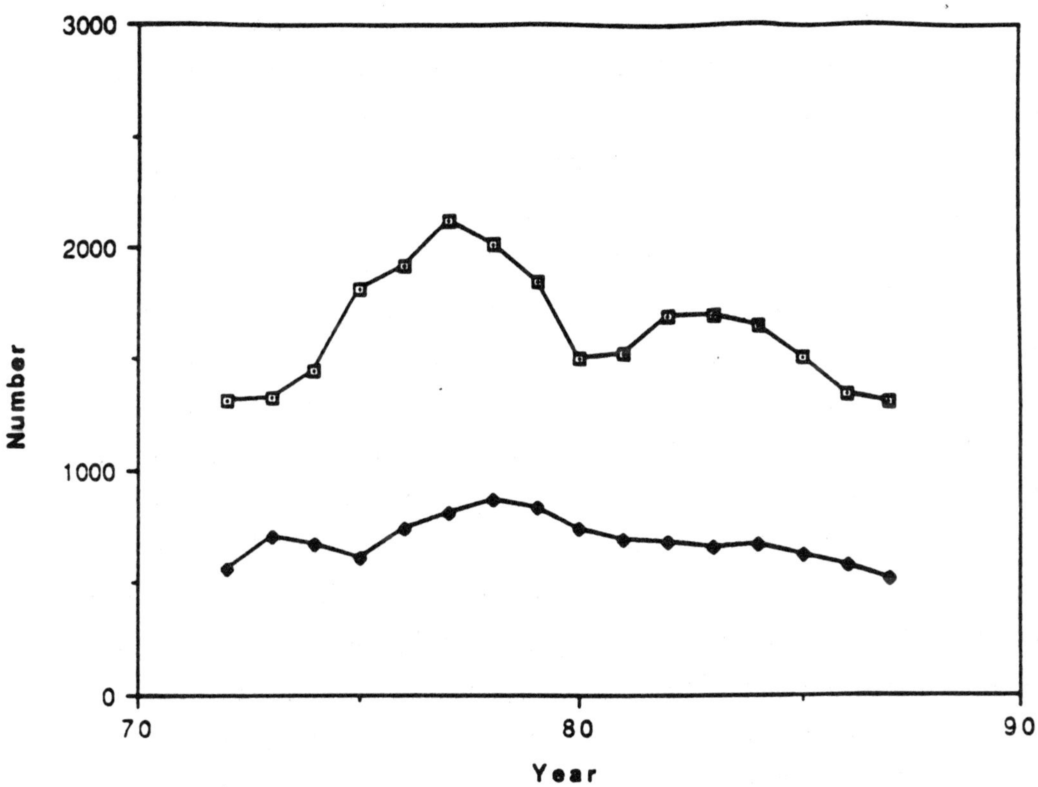

Figure 1. Nuclear engineering undergraduate enrollment and degrees, 1972-1987.

The industry and university response to the TMI accident was to reexamine safety issues and the requirements for training of reactor operation and maintenance personnel. The university response included an increased curricular emphasis on thermal transport in reactors as well as safety and licensing issues (Argonne Universities Association, 1980). The national industrial response was to create an industry entity, the Institute for Nuclear Power Operations (INPO) which established national standards for training requirements to support reactor operations and accredited the training programs of individual utilities. Similar efforts were undertaken at an international level by the IAEA.[3]

The Chernobyl nuclear reactor accident in the Soviet Union in the early '80s caused additional negative pressure on nuclear engineering enrollments. Delays due to public intervention in the licensing process led to substantial cost overruns. High capital expenditure rates resulted from the high interest rates of the late '70s and the lack of income during the construction and licensing phases. Many commercial nuclear power plants that had been planned in the '60s and early '70s were cancelled by the mid '80s. Public perception of nuclear power continued to be negative, with uncertainties in the risk due to accidents and the long-term storage of nuclear waste stressed. Nevertheless, the United States presently has more than 100 commercial nuclear power plants in operation, which generate approximately 20 percent of the its electrical energy supply.

Contrary to common perception, the demand for nuclear engineers has remained strong during the past two decades. The B.S. nuclear engineer has not experienced the period of slack employment typical of the aerospace and petroleum industries in recent years. Indeed, many utilities and vendors have been forced to use engineers with only a nuclear engineering minor or without any specific nuclear training to fulfill their nuclear engineering functions. The job prospects for nuclear engineers continue to be strong (Basta, 1988). There is growing work for nuclear engineers in the areas of environmental restoration and waste management. There is an expanding international market in nuclear power, and the U.S. nuclear industry is gearing up to provide nuclear services to that market. The U.S. government's defense-related nuclear engineering work has continued to grow. The growth in management, technical, and professional positions in nuclear-related industries is expected to be at an annual rate of 2.3 percent for the next decade, but the expected annual growth in nuclear and reactor engineering employment will be over 8 percent (DuTemple and Diekman, 1988).

---

[3]This resulted in the publication of additional guidelines, such as *Manpower Development for Nuclear Power, A Guidebook*, Vienna, Austria: IAEA Publishers, 1980.

Future of Nuclear Engineering

The nuclear engineer (NE) of today has a core training in the fundamental engineering sciences. Thus he or she has the flexibility to function in many standard non-nuclear engineering jobs such as project management, engineering design, systems analysis, and operations. Training in the fundamentals of engineering science, along with synthesis and engineering design, produces a professional engineer with the lifelong ability to acquire new technologies and change career direction. Because of a strong foundation in the interaction of radiation with matter and in nuclear processes the NE has a role in research and development in areas of emerging nuclear technologies, including the use of controlled nuclear fusion as an energy source and the recently touted cold fusion research. Additionally, the use of nuclear fission reactors as non-electrical energy sources in various environments and the use of radiation in material and biological systems are emerging areas for nuclear engineering. Shielding calculations for the use of accelerators in medical treatment environments and radiation health physics are also applications of nuclear engineering.

The nuclear engineering curriculum could be used as a model for the development of an inherently flexible engineering training program (Table 3). Each engineer should receive two or three years of training in the fundamental sciences and engineering sciences, as well as the social sciences and humanities, to provide appropriate breadth. That would be followed by one year of specialization in a particular discipline (i.e., nuclear). The final year would consist of a practicum in engineering practice and design, common to all disciplines, but with specific applications in the chosen major. The advantage of a curriculum that emphasizes the major area only in one year of study is the subsequent flexibility to change to another engineering discipline with a single year of study spread over a number of years in a part-time format, while one is employed full-time as an engineer.

The acceptance of such a general curriculum by a broad spectrum of engineering disciplines is unlikely because of growing demands placed on each curriculum by the march of technology in the individual discipline. The typical B.S. engineering program in the United States is now a 4-5 year program, due to the demands for breadth and depth in the course work. Curriculum pressures push the engineering student to a decision on engineering major early in the academic career. However, some programs have continued to maintain a commonality of the engineering curriculum through the sophomore year. The

Table 3. Proposed General Engineering Curriculum

Freshman Year:	Science, math, humanities and social sciences including physics through modern physics, chemistry through quantitative analysis, mathematics through partial differential equations and numerical methods, social sciences and humanities, english composition and public speaking, economics, other electives.
Sophomore Year:	Engineering sciences, including engineering materials, engineering economics, heat transfer, fluids, statics, electronics, thermodynamics, computer sciences, laboratory experimentation, advanced sciences.
Junior Year:	Emphasis on core of discipline; for nuclear engineering this would include radiation interaction with matter, neutron diffusion and kinetics, reactor thermal hydraulics, reactor safety and licensing, radiation health physics and shielding, nuclear instrumentation and reactor laboratories.
Senior Year:	Practicum in professional engineering practice, engineering design and industrial practice.

acceptance of a common curriculum in the final year is the hurdle that must be overcome to make the program outlined above feasible.

Many nuclear engineering academics agree that nuclear engineering as a discipline is, and should remain, independent of the commercial nuclear power industry—just as electrical engineering as a discipline is independent of the electrical power and microelectronics industries and mechanical engineering is independent of the machine design and automotive industries. While many NEs find employment with nuclear utilities, others find employment with national research laboratories, with state government, with the aerospace industries, or in university teaching or research. Nuclear engineering is flexible enough that one can enter it from the graduate degree level or from an undergraduate option level, as well as the now traditional B.S. curriculum.

Nuclear engineering continues to grow and mature as a discipline. With the evolution of the nuclear power industry and the continued development of applications of radiation, it will continue to do so well into the twenty-first century. The present hiatus in construction of commercial nuclear power plants notwithstanding, there continues to be a demand for nuclear engineers in the U.S. marketplace. The goal of nuclear engineering educators must be to continue to produce a product that is based in the fundamentals of nuclear processes and applications, with an underpinning of the core of engineering science fundamentals.

References

Argonne Universities Association (AUA), Nuclear Engineering Education Committee. 1980. *Education and Training Needs of the Nuclear Power Industry*. Argonne, Ill.: AUA.

Basta, Nicholas. 1988. Job prospects for computer engineers. *Graduating Engineer* 9(4:March):72-73.

Dawson, Frank G. 1976. *Nuclear Power, Development and Management of a Technology*. Seattle: University of Washington Press.

DuTemple, Octave J., and W. M. Diekman. 1988. *Job Opportunities in the 1990s*. La Grange Park, Ill.: American Nuclear Society.

construction of commercial nuclear power plants notwithstanding, there continues to be a demand for nuclear engineers in the U.S. marketplace. The goal of nuclear engineering educators must be to continue to produce a product that is based in the fundamentals of nuclear processes and applications, with an underpinning of the rest of engineering science fundamentals.

References

Argonne Universities Association (AUA), Nuclear Engineering Education Committee. 1986. *Education and Training Needs of the Nuclear Power Industry*. Argonne, Ill.: AUA.

Basta, Nicholas. 1988. Job prospects for computer engineers. *Graduating Engineer* 9(4; March):72-75.

Dawson, Frank G. 1976. *Nuclear Power: Development and Management of a Technology*. Seattle: University of Washington Press.

DuTemple, Octave J., and W. M. Goldman. 1988. *Job Opportunities in the 1990s*. La Grange Park, Ill.: American Nuclear Society.

ADAPTING TO COMPUTER SCIENCE

Saul Gorn
University of Pennsylvania

Introduction

Although I am not an engineer who adapted himself to computer science but a mathematician who did so, I am familiar enough with the development, concepts, and activities of this new discipline to venture an opinion of what must be adapted to in it.

"Computer and information science," known as "informatics" on the European continent, was born as a distinct discipline barely a generation ago. As a fresh young discipline, it is an effervescent mixture of formal theory, empirical applications, and pragmatic design. Mathematics was just such an effervescent mixture in western culture from the renaissance to the middle of the twentieth century. It was then that the dynamic effect of high speed, electronic, general purpose computers accelerated the generalization of the meaning of the word "computation." This caused the early computer science to recruit not only mathematicians but also philosophers (especially logicians), linguists, psychologists, and economists, as well as physicists and a variety of engineers.

Thus we are, perforce, discussing the changes and adaptations of individuals to disciplines and especially of people in one discipline to another. As we all know, the very word "discipline" indicates that there is an initial special effort by an individual to force himself or herself to change—to adapt one's perceptions to a special way of viewing certain aspects of the world and one's behavior in order to produce special results. For example, we are familiar with the enormous prosthetic devices that physicists have added to their natural sensors and perceptors in order to perceive minute particles and to smash atoms in order to do so (at, we might add, enormous expense and enormous stretching of computational activity). We are also familiar with the enormously intricate prosthetic devices mathematicians added to their computational effectors, the general symbol manipulators called computers.

The disciplines require us, perhaps less dramatically but always more subtly, to change the way we understand, how we act, how we act in order to understand, what we should understand in order to act, etc. The adaptation of a discipline over a period of time is described by its history; this is sometimes slow, in the human scale, over the millennia, and sometimes so fast for human beings as to be considered revolutionary, as in our two examples, or, more crucially, in the scientific revolutions whose structure was studied by Thomas S. Kuhn (1970).

Because of this concern with adaptability of individuals and disciplines, it is useful to consider a rough taxonomic subdivision of disciplinary attitudes. Just as the individual's perception and behavior is a result of the flow of signals, one way from receptors to perceive and the other of commands to effectors to act, so are there two types of disciplinary signals; there are descriptions of aspects of the world, presented in declarative sentences, and there are prescriptions of what ought to be done about those aspects, in imperative sentences. Theories are logically or empirically organized systems of the former, and programs are organized systems of the latter. Some knowledge-oriented disciplines are mainly concerned with producing theories in what we might call their "object language," reserving their programmatic statements to the discussion of their methodology, in what we might call their "meta-language"; examples are pure mathematics (where the methodological language is that of logical programs) and the empirical sciences (where the methodological language is that of experimental programs). Other disciplines are mainly "action-oriented" and concerned with programs such as hunting, cooking, manufacturing, and constructing. Still other disciplines—such as law, medicine, and engineering—have as their purpose applying knowledge to achieve action. These last must be sensitive to both the changes in knowledge and the changes in activities; they therefore change more than the knowledge-oriented and the action-oriented disciplines they are bridging; and if one of these goes through a revolution, the connecting professional discipline may well follow suit. Changes in a profession caused by changes in the disciplines bridged form the "professional change syndrome."

A young discipline, like a young person, is an effervescent mixture of all three of the above-mentioned types of activity. Until the mid-twentieth century, mathematics and physics were like this. Computer science is like this now.

As time goes on, the action of perceiving for the discipline may disturb what is being perceived, especially if the perceiving and perceived elements are the same order of the magnitude; this was the case with particle physics and quantum mechanics, resulting in the indeterminacy principle. More generally, the knowledge-oriented and action-oriented aspects of the young discipline will conflict, putting the discipline through a crisis and possibly a revolution. This has happened to mathematics. Engineering must adapt to both the change in mathematics and the change in physics, two examples of the professional syndrome in engineering.

Since the success of engineering—or engineering aspects of mathematics, physics, and computer science—is measured by how many problems it solves, there is a further cause of change, what might be called the "solved problem syndrome." Here the result may be a standard technique that no longer belongs to the profession but is either relegated to the discipline that presented the action problem—for example, all of applied mathematics—or becomes a new separate discipline. In the case of engineering, an old example of the latter is plumbing. In transportation such results have been the train-"engineers", flight-"engineer", etc. Somewhat similar is the engineering application of x-rays and electro-magnetics to medicine, resulting in radiology. The latter is still changing, having developed CAT-scans, PET-scans, magnetic resonance techniques, and, by another application of computer techniques, three-dimensional graphic imaging.

To see how engineers must adapt to computer science, we will first see how the engineering aspect of mathematics was originally computational. Then we will see how those computational aspects of mathematics developed to involve not only physical engineering but also psycho-linguistic aspects. This will lead us to enumerate the areas of computer science, each of which alone has a formal, an empirical, and a design aspect. Some of the formal aspects will be novel to engineers, adding to the mathematical background a computer engineer requires beyond the analytic background all engineers need; but all the design aspects will of course be engineering-oriented. What is more, so many of these design aspects will be not physical but, rather, psycho-linguistic and, hence, close enough to symbolic and human activities to be most novel to engineers. These will require the greatest adaptation.

How Mathematics Separated into Pure, Applied, and Computational

Because mathematics arose from the shepherd's need to count, the agricultural need for measurement and astronomical information, and the urban commercial need for both, calculation and geometric construction techniques are, of course, ancient. That summation of Pythagorean mathematics, the "elements" of Euclid, therefore contained those primitive programs, the constructions with straight-edge and compass, and those sophisticated verifications of the correctness of those programs, the formal geometric proofs. Euclid's students could therefore inscribe in a circle regular polygons with 3, 4, 5, 6, 8, 10, 12, 15, 16, 20, etc., sides, thereby computing lengths—considerably more than agriculture required. Meanwhile the Greek, and later the Roman, worlds made mechanical use of counting boards (abaci) and counters (calculi) to do their counting and sums.

However, it was not until the eighth century A.D. that the western world could begin to profit from the Hindu programming of numerical computation. It was then that Al Khowarizmi in the Arabic world made specific the algorithms of algebra; and it was only several hundred years later that the Arabs picked up the Hindu numeral notation itself. It was therefore only in the renaissance that the numerical symbolism and its resulting algorithms came to be fully developed. But it already had gone beyond simple numerical calculation into algebraic and trigonometric calculations—for astronomy as well as for commerce. Even so, the Arabic-Hindu numerical data structures and the corresponding algorithmic programs for addition and multiplication did not come into common use until the eighteenth century!

Thus, from the renaissance until the mid-twentieth century, at a leisurely but slowly accelerating pace, mathematicians kept expanding their symbolic notations via their algebraic expressions, number theoretic expressions and functional expressions, and computed expressions that were formal derivatives or exact solutions in terms of the elementary functions to differential and other functional equations. They were expanding their language of descriptive, implicit and prescriptive, explicitly programmed specifications. Solving equational problems meant converting from the first type (declarative) to the second (imperative) specification languages. Note that this was the reverse of the Euclidean verification of the imperative construction procedure by the declarative language of formal proofs mentioned before.

At the same time that mathematicians were doing notational and computational innovation, they were simultaneously developing informal theory and applying both to physics and astronomy (these were the three types of activity mentioned in the introduction). To name only a few in this development over the centuries, there were Vieta, Fermat, Descartes, Pascal, Newton, Leibnitz, Euler, the Bernouillis, Lagrange, Laplace, Cauchy, and Gauss. Interested not only in pure mathematics—the fundamental theorem of algebra, number theory, and differential geometry—but also in applied mathematics (astronomy and physics) and in symbolic invention and new extensions to computation—the congruence notation in number theory, algorithms for solving systems of linear equations, the method of least squares, and error statistics—Gauss was a formidable calculator and programmer. In fact, it was his early developed theory of ruler and compass construction, his specification of exactly which regular polygons could be so constructed, and his program for doing it specifically for the regular polygon with 17 sides (when he was 19 years old) that made him decide to become a mathematician rather than a linguist.

Some of Gauss's predecessors had also been famous as philosophers—Descartes, Pascal, and Leibnitz, for example—concerned with logic and the theory and practice of computation. For example, Leibnitz, on the one hand, invented the differential quotient and integral notations and informally derived the computational rules of the differential and integral calculus. On the other hand, he described a formal algebra of logical "and," "or," and "not" as well as predicted the future formal logical algebra that he called a "universal characteristic."

In any event, by the end of the eighteenth century the informal concepts of "infinitesimal," "infinite summation," "continuity," and "differentiability" became disturbingly paradoxical; and yet they had such fertile applications to physics that something had to be done to verify what could be verified in the symbolic procedures and to formalize them so as to safeguard them from paradoxical misuse. The whole nineteenth century was occupied with this goal.

This kind of problem was not new. The Pythagoreans had such great success in their reduction of measurements to counting, by the use of fractions, that they were confounded by the seeming paradox that the diagonal of a square could not be "rationally" related to the use of its side as a unit. The solution of this early mathematical crisis was the geometric theory of ratio and proportion and its application to the theory of similar triangles as given in the tenth book of Euclid; neater construction theories of the "real numbers" had to wait until the theories of Cantor, Dedekind, and Weierstrass in the later half of the nineteenth century.

The Pythagorean failure of the possibility of measuring by simple counting was only partly solved by the invention of fractions and real numbers. There still remained logical confusion in language; for example, Zeno's paradox of Achilles and the Tortoise not only concerned itself with an infinite series having a finite limit, but also muddied the water by setting up a confusing correlation of a particular linguistic description of the race between the two with the actual occurrence—the time taken to describe the race exactly had nothing to do with the time taken to run it. The nineteenth century development of a formal logic even more severe than those employed by Aristotle and Euclid brought out more mathematical crises and logical paradoxes.

For example, Boole in the early nineteenth century developed a symbolic calculus that he called "laws of thought." One such law, called "modus ponens," states that if we have two declarative sentences, call them p and q, and if we know that the sentence p is true and also that the sentence of the form "if p is true, then so is q" is also true, then we can conclude that the sentence q is also true. Boole had a neat algebraic symbolization of this procedure, an algebraic expression: [p& (p->q) ->q]. This is, however, a functional expression in a Leibnitzian calculus of two-valued logic; it has the value T (for "true") for each of the four possible combinations of values T and F (for "false") that p and q can have. Such a function is called a tautology (i.e., always true). Lewis Carroll at the end of the century, in a paper called "What the Tortoise said to Achilles," pointed out that to achieve the result of modus ponens would require not merely this tautology but also one that states that p and p->q, and it, [p& (p->) ->q] = A_1, imply q—that is,

$$A_2 = [p\& \ (p\text{->}q) \ \& \ A_1 \ \text{->}q],$$

and further that

$$A_3 = [p\& \ (p\text{->}q) \ \& \ A_1 \ \& \ A_2 \ \text{->}q], \text{ etc.}$$

In other words, it takes an infinite number of these tautologies to achieve modus ponens.

The law of thought, modus ponens, is not in the Boolean expression language; rather, it is a valid procedure in a prescriptive meta-language that talks about the production of proofs in the Boolean object language. The tautologous symbolic expression had been designed to symbolize modus ponens, and the symbol was confused with what it symbolized. The tautology is no more a law of thought than the artist Magritte's picture of a pipe, which he entitled "this is not a pipe," is a pipe!

In general, crises in mathematics were signaled by paradoxes. The paradoxes were usually resolved by finally recognizing that they were either (1) reductio ad absurdum proofs that something did not exist or (2) reductio ad absurdum proofs that some goal was impossible to achieve by a finite means, or that mixing meta-language with object language caused us to confuse symbols with what was being symbolized. And even in the reductio ad absurdum solutions, the main problem often was to find out what was presumed to exist but did not, or why it did not. For example, "the real number 2 is not rational" or "the number of prime numbers is not finite."

Many of the dramatic advances were due to lifting the level of thinking by meta-linguistic description, and by generalization at the meta-language level. For example, algebraic symbolization is a generalization into meta-language of the object language of arithmetic—and analytic symbolization is a generalization into meta-language of the algebraic language of arithmetic functions (not all functions are algebraic or even representable by infinite series of algebraic expressions). But the symbolic procedures suggested at meta-language symbol level might not have valid meaning in the operations at object language level. The mixture of object language and meta-language is intuitively powerful but, like most powerful tools, is dangerously capable of misuse. This was why mathematics went even further than Euclid in carefully formalizing proofs. Archimedes and Euler intuitively handled infinite series correctly, but too many paradoxical operations, such as hidden divisions by zero in algebraic meta-language or its equivalent in continuity and differentiability discussions, made careful logic necessary.

And then the paradoxes even turned up in the logic! Cantor, for example, gave a formal definition for two sets having the "same number of elements"; it was later used to give a formal definition of cardinal and ordinal numbers, even for infinite sets. He was able to show that there were just as many rational numbers between zero and one as there were natural numbers all told (i.e., that the rational numbers are countably infinite), a seeming paradox but not really one; and he capped this result with a reductio ad absurdum proof by his "diagonal method" to show that the number of real numbers between zero and one is not countably infinite (i.e., has a larger infinity than the number of natural numbers).

All this activity of formalization to avoid paradoxes and contradictions was resulting in an extension of the Pythagorean program of reducing all mathematics to numbers; it was all being reduced to a formal theory of sets and then (by Frege and later by Whitehead and Russell) to formal logic itself, as foreseen by Leibnitz. But, the logic itself emerged with dangerous paradoxes, beyond what Lewis Carroll had recently found in Boole's laws of thought. All this thinking about thinking and mixing meta-language with object languages seemed fraught with dangers due to self-referencing: the numbers of sets of numbers, sets of sets, etc.

In fact Russell confounded Frege, immediately after Frege's publication with his famous paradox of "The set of all sets that are not members of themselves" (is it or is it not a member of itself ?!?!). And some further paradoxes brought to the surface the relationships to language. For example, there was the Richard paradox, which first pointed out that there could be only a countably infinite number of English phrases that specify functions of one natural number variable and then used Cantor's diagonal argument to show that this could not be true. And then there was Berry's paradox: "The least natural number not nameable in fewer than 22 syllables" is hereby named in 21 syllables. Thus, by the beginning of the twentieth century, a larger and larger portion of mathematical activity was the axiomatization and presentation of mathematical theory in an even more rigid formalization than Euclid's elements. Called "pure mathematics." by Hilbert, it axiomatized and presented a formal system that would cover all possible mathematical truths; any properly (i.e., mathematically) stated sentence in such a system would be formally provable or else formally contradictable. This optimistic program was formally proved to be unattainable toward the middle of the century; Gödel showed that any sufficiently rich formal system was either inconsistent or incomplete in the sense that there would be statements within it that were undecidable (i.e., could neither be proved nor disproved). His proof was something like Richard's paradox, using a Cantorian diagonal argument on a formalization of a meta-language. About the same time, logicians were uncovering a number of undecidable questions and unsolvable problems by developing a number of equivalent theories of computation.

All through the history of mathematics, the word "computation" had been expanding its meaning. The formalization process of the nineteenth century accelerated it by including all the algebras, geometries, and logics. But it was the necessity to formalize the programming of computations caused by the inhuman speeds of the recent general purpose computers that made computation include all precisely specifiable symbol manipulation. Now general symbol manipulation includes the pragmatic effects of deliberate ambiguity caused by shifting interpretation—for example, between object language and meta-language. General computer programming must therefore include what mathematical logic programming must forbid. Again, the mathematical study of arithmetic is independent of the way numbers are represented; the computer program for addition depends not only on the particular computer type, but also on the number representation system used. There are so many pragmatic questions in computer theory, use, and design that the area of computation and the area of mathematics were pulled apart.

For a different reason, mainly what we called the "solved problem syndrome," applied mathematics also separated from pure mathematics.

By the middle of the twentieth century, the applications of mathematics had spread to so many different disciplines that the spread was too vast for any single person's curriculum. The mathematical applications, once the mathematical aspect of their disciplines had been precisely specified and handled, were relegated to the discipline of the application. A few universities were concerned with a united applied mathematics group but, by and large, the discipline became restricted to "pure mathematics."

How Computation Broadened its Scope and Developed a Close Relationship with Engineering, Logical Theory, and Psycho-linguistics

Mechanical aids for counting became a necessity especially in urban cultures—for example, around the Mediterranean in Hellenic and Roman times. And from those beginnings computation, like mathematics generally, developed in three interacting strands:

(1) The development of the physical aids and their psycho-linguistic aspects into machines; the physical aids development is the obvious engineering aspect, the psycho-linguistic the more subtle one;

(2) The theoretical aspect—beginning with mathematics as a whole, but narrowing down to formal proofs, and then to formal logic itself; and

(3) Forming a spiral, the psycho-linguistic development of naming and specifying the data structures (at first, the number representation systems) and the expressions of the procedures for handling them (their programs or algorithms).

During the interacting development of these three phases in computation, it was the expressions of the procedures that were turned into machines. For example, as mentioned before, in Hellenic and Roman times, counting and adding were facilitated by counting boards with lines on them for placing counters. These static devices, made mobile by fleetly calculating human fingers, were of two designs; the first, a simpler design but more bulky, had room for 10 counters per line, while the second, a more compact one, had 5 counters per line, and intermediate lines designed for at most 2 counters to indicate the number of full hands of five fingers that had accumulated on the base lines. The description of the counting and adding procedures for the bulky counting board was, of course, much simpler. The two types of Roman and Chinese abaci are further developments of these placid machines, where the lines are replaced first by slots and then by wires, and the counters are beaded in or on them vertically instead of being strung out horizontally; in the more compact type the intermediate lines are vertically above the ones they relate to.

The Roman symbolization of the total number counted on a board was a compact description of the picture on a compact board, using I and V for the first line and its associate, X and L for the second, C and D for the third, etc., and a reversing grammar to obtain extra compactness, IV for IIII, XL for XXXX, etc. The parsing of these number representation phrases was fairly complicated in order to achieve compactness and representation on one line.

The Hindu-Arabic symbolization—especially when a symbol for an empty line, zero, was introduced—was a compact description of the bulkier abacus. This second phrase structure language used the position of the numeral to represent the position of the wire and thereby avoided the necessity of using a larger and larger alphabet for more and more wires. Moreover, the algorithm for addition in this language had a considerably simpler description than did that for addition in the Roman numeral phrase structure language.[1]

It would seem that it took mankind 1,000 years to decide that the descriptive language for the bulkier machine was more useful. The algorithm for addition in the Arabic symbol system has a much simpler description as well as a need for only a finite alphabet.

The 19-year-old Pascal assisted his tax-collecting father by designing a more mobile mechanical adding machine than the abacus. It simulated in hardware the program for addition in the Arabic system. And within a generation Leibnitz adapted the same technique to make the machine do multiplication, albeit with the vigorous help of the human hand (i.e., some software was still necessary).

The explosion of functional equation-solving techniques in mathematics during the next 200 years made a more Gauss-like calculator desirable. At the beginning of the nineteenth century, Charles Babbage felt that all function table construction should be made mechanically and automatically, and he spent some decades designing a "difference engine" that would do the job. He was also convinced that a device similar to the cards controlling a Jacquard loom in textile weaving could be used to allow the descriptions of many procedures to serve as different programs in one and the same "analytic engine." Although his mechanization ambitions were defeated because technology was not advanced enough to standardize the necessary parts, these ambitions were prophetic enough to anticipate Hollerith's punched card machines at the end of the century and the electro-mechanical relay machines of Konrad Zuse, Howard Aiken, and George Stibitz 100 years after his time. In effect, Babbage had invented the concept of programming general purpose

---

[1]Because the natural numbers form a simple chain, the same linguistic expressions, the number namers, serve the double purpose of being enumerators—while the counting proceeds—and summarizers—when counting ends. When we want to stress this difference, we use ordinal language instead of cardinals; their isomorphism is a deep property of finite natural enumeration. This isomorphism is lost in both infinite cardinals versus ordinals and in "tree names" as against "tree addresses" in ramified structures such as decision trees, family trees, classification systems, or organization charts.

calculating machines and the idea of programming languages for the specification of how to run numerical algorithms. His message was kept alive not only by the designers of machines in the 1940s and '50s of this century in the United States and Britain and onto the European continent, including the Russian mathematicians and engineers following Tchebycheff, but also by the logicians in the 1930s, '40s, and '50s. These latter developed a general theory of computation even before electronics took over.

We therefore return to the portion of the history of mathematics where an extension of the concept of computation was occurring in the nineteenth century. We have already noted that following Al Khowarizmi a meta-language for arithmetic, namely algebra, appeared and flourished. The solution of algebraic equations as explicit functions of their coefficients was sought. Where the solutions were impossible in the number systems already available, new extensions to the number concept were invented. We have already remarked on this happening when the Pythagoreans had to deal with that anachronism of the future, $x^2 = 2$. Then, through the Renaissance, it resulted in the invention of the next imaginary class, the negative numbers, yielding the solution to all equations of the form $ax + b = 0$, except, of course, for $a = 0$. Then a unified solution to all quadratic equations—$ax^2 + bx + c = 0$—was found as a function of a, b, and c, provided the invention of numbers was extended to include another imaginary class, first called imaginary numbers and finally accepted as the "complex numbers"; and the unified solution called for addition, subtraction, multiplication, division, and, the Pythagorean crisis having long been understood (?), the operation of taking a square root. The program, or algorithm for the construction of the solutions, was presented explicitly in an algebraic form. What could be said about all algebraic equations? Gauss finally proved that the(complex numbers would suffice to solve them, but not by an algebraic formula. (We would now say that "the complex number system is algebraically closed.") By his time the construction of the solution by addition, subtraction, multiplication, division, and the extraction of roots had been achieved for degrees 3 and 4, and no more.

It was then that Galois, by studying the effect of permuting the roots of an equation, discussed the algebraic systems called groups. He found that those and only those equations were "solvable by radicals" whose Galois group has a certain restricted structure; and he further found that the general algebraic equation of degree greater than or equal to five did not have this structure, called, incidentally, "solvability." Thus another condition for solvability and unsolvability emerged, as had those that prompted the invention of extensions to the number system, or Gauss's condition for solvability by straight-edge and compass construction.

Computation now included combinatoric ones concerned with rearrangements of objects and counting arrangements fulfilling various conditions. The theory of probability had already depended upon such combinatoric processes in the study of games of chance, at least since the time of Fermat and Pascal. And Euler had considered counting and classifying paths on road maps or on edges of polyhedra in the beginnings of combinatory topology known as "analysis situs." The specification of the new algebraic groups themselves appeared in a study of strings of characters, each character representing a generator, where certain pairs of strings were identified as producing the same group elements. Computation now included the purely syntactic study of certain types of symbol manipulation (ignoring the symbolism); its application to group theory and a variety of algebras, geometries, and topologies were studied through the nineteenth century by, for example, Hamilton, Cayley, Poincare, and Burnside. At the outbreak of World War I, Thue was publishing papers on "word problems" that were forerunners of some of the general theories of computation. Cantor's diagonalization proof was a syntactic game with number representation words. The logical paradoxes were semantic games with words as well. Logicians of the Polish school, their student Herbrand, and Gentzen were also thinking in terms of syntactics and semantics of such symbol manipulation games. Finally, after Gödel's work mentioned above, the logicians of the 1930s and '40s followed three types of approach (and their mixtures) to a general theory of computation. They all produced important examples of unsolvable problems, showing the limits of such general computation; and the theories were proved to be equivalent.

The first approach presented purely syntactic "rewriting systems" as in the "presentations" of groups, and the generalization toward word problems by Thue. Post's "production systems" or "semi-Thue" systems was an example, and Post presented the undecidable "correspondence problem." Some years later it was shown that the word problem for semi-groups and groups, namely determining whether two products of generators represented the same element, was undecidable. In this class, some years later, was Markov's theory of algorithms.

The second approach presented formalizations of the meta-language of such syntactical symbol manipulation; the characters in the meta-language had meanings (semantic content) and were therefore really symbols, where the object language characters might not be. For example, Church, in his λ calculus, eliminated the confusion in mathematical notation between symbolization of a function and symbolization of the value of a function for a general element in its domain. Rosser concerned himself with the possibility of reducing a λ expression to a "normal form." Curry considered primitive symbol manipulating operators, the "combinators", and their combinations. And Kleene formalized recursion to produce a theory of recursive functions. This approach has had a direct effect in recent studies of programming language semantics and in the theory of program verification.

The third approach was more pragmatic in that it specified the users or interpreters of the symbols being manipulated as well as the syntactics and semantics of the manipulations. These users and interpreters were the symbol manipulating machines, and the approach was directly descended from Babbage. The prime example of this approach was that of Turing. He specified the concept, that of Turing machine, as any mechanism that had a finite number of states and a potentially infinite tape that, at each cell, could be read, erased, printed on, or switched to another cell depending on the state the machine was in, and had a switch to another state as part of the interpretation of the symbol. Prime illustrations of Turing machines were concatenators of strings of symbols, copiers, and combiners. A carefully designed combination of a few of these Turing showed to be universal in the sense that it could copy the specification of any special machine and imitate it. The tapes specifying Turing machines served as programs for the universal Turing machine. The concepts of program, procedure, algorithm, and machine had become identified as far as their effect was concerned, even though their uses of space and time might be quite different. Turing showed that it was impossible to construct a program that would accept any program and appropriate data and determine in a finite time whether or not the problem represented by the program and data would come to a halt or would continue indefinitely. During World War II Turing was also involved with the design and use of a machine in Britain, the Colossus, whose main purpose seems to have been cryptoanalytic.

Next developed were general purpose machines and electronic developments. The electro-mechanical calculating machines were already so much faster than human computation that they required prior program preparation. Rutishauser even contemplated making the programming process itself partly automatic, foreseeing the extension of the meaning of the word "computation" that the succession of generations of machines were about to add to mathematical general symbol manipulation, namely the construction of computer programs themselves. The first electronic, general purpose calculating machine, the ENIAC, was at first programmed for special problems by hand-plugging the interconnection of the registers from which and into which data were to be moved and operated on. Later the programs were coded numerically.

The EDVAC, immediately after the ENIAC, was designed with a fast-access internal storage (mercury delay lines) of 1,024 words that could be either data or coded instructions. Effectively, the program was the switching design of the special purpose machine one wanted the general purpose machine to become; contents of a storage cell were data if they were sent to an adder, say, but were an instruction if they were sent to the "instruction register" to be interpreted (i.e., decoded, or parsed) as an instruction. Not infrequently, in a programmed loop, the first was done to change the address of the addend to be used, and immediately afterwards the second was done to the result. The machine was modifying its own instructions, as though self-consciously. This brings to mind the mixing of meta-language (i.e., the programmed instructions) and object language (the computational data).

On the one hand, this common storage of instructions and data led to the construction of programmed compilers and interpreters of higher level languages; on the other hand, it illustrates the principle of the "logical equivalence of hardware and software." It was later modified by the introduction of standard binary codes for the complete typewriter keyboard. The data were no longer viewed as just numerical. The similarity of these electronic symbol manipulators to the universal Turing machine was now more obvious than ever.

The programming of interpreters of higher level languages now forced the self-conscious examination and imitation of psycho-linguistic processes. Meanwhile, mathematics was being applied to psychology and linguistics. Mathematical psychology at first concerned itself with stochastic learning models of the reinforcement type; its only effect on computation was to sharpen the programming technique of delayed random selection. In fact it was the attempt to program problem solving with its unpredictable storage requirements, and to do it on a machine with magnetic drum storage, that prompted the invention of list structure and "push-down" storage techniques in this country. This led to dynamic storage allocation and advanced the development of the area called artificial intelligence. Mathematical linguistics introduced the Chomsky hierarchy of phrase structure languages (Luce et al., 1963). These were approximations of the linguistic procedures employed by human beings in their use of natural languages, but immediately enriched the theory of computer languages. They gave us the various types of automata, finite state, and push down, needed to parse them and models of formal languages (e.g., finite state and context-free) useful in studying and designing computer languages and analyzing the general computation process.

At first the coding of this computation process was done by specialists—the machine coders; in addition, there were the specialists (sometimes these same machine coders) who entered the problems in the machine, ran them, and arranged for their reading and printing. The machine time was too expensive to be slowed down by the single user; the set of problems of a number of users was batched and scheduled by the operators.

The advancement in machine technology came at a rapid rate, though, luckily, at a slow enough rate so that internal, fast access storage was still expensive and therefore too small to afford separating instructions from data. Thus, when the magnetic auxiliary storage devices, tapes and drums were enhanced by the faster access, internal storage (the core memories), the internal storage could increase moderately. The increase was enough to permit the specialized coder to be replaced by a program, the aforementioned compiler or interpreter of some standard higher level language; it also permitted a partial replacement of the machine operator by an automatic scheduler that shared the time allotted to a number of users, their inputs, their required auxiliary storage, and their outputs. This latter software was an "operating system" that also handled the allocation of internal storage dynamically, not only to schedule the time sharing of the users but also to allocate unpredictable storage needs of the single user. Machine users needed at least two types of dynamic storage allocation: (1) recursively defined functions in such higher-level programming languages as ALGOL and LISP and (2) the unpredictable storage necessary in the automatic simulation of human problem solving in the new area of artificial intelligence. Thus the human calculating processes, even at the programming and operating level, were self-consciously examined and programmed. The programming and operating level itself was recognized as a type of calculation much like the algebraic, analytic, and symbol manipulative level.

Meanwhile, the extension of programming capability to the machine user caused the explosion of computer applications to all disciplines sophisticated enough to develop advanced symbolic manipulation—including even such action-oriented disciplines as sports, with their diagrammatic simulation representable in computer graphics. The market expanded rapidly and supported the hardware, software, and application research. The machines advanced in generations of 5-10 years—the transistorized generation, semi-conductor devices, large-scale integration, and the VLSI methods now used to construct microprocessing chips the size of fingernails. Simultaneous with the drastic decrease in size came a drastic increase in speed, a drastic lowering of cost, and a drastic expansion of the market because of a drastic extension of the meaning of computation and its applications. For example, business data processing applications were once again, as in the Renaissance, an important use of machines, but now with the counters, recorders, and summarizers (i.e., the large staffs of clerks) also replaced by machines.

Meanwhile, during one generation of engineers and clerical staffs, the hardware and software went through three or four generations of change. What is more, the software production is now much more expensive than the hardware and needs fully as much engineering. Far from the secrecy about computational design and use that characterized the international and intra-national relations immediately following World War II it is now clearly more profitable for all to standardize types of machine design, language design, and information interchange. The computer systems and computations, therefore, are not merely proliferating; they are actually clustering into larger and larger networks for the interchange of such computed information. It is not merely airlines that need such "distributed system computation," but also all extended business corporations and international academic research.

In summary, the word "computation" now means "any precisely specified process of symbol manipulation," even pictorial. When restricted to computer programming, the semantics (i.e., the meanings of symbols) is restricted to the symbol manipulative. But these meanings can be expanded to visual and other prosthetic sensing and effecting devices we choose to add to the machine's data assemblers and output reactors. This is what is done in the computations for robotics. And, indeed, some of the most humanly interactive programming techniques are now object-oriented, mixing digital video graphics and digital audio sound, in distributed computation and information networks.

The Areas of the Discipline of Computation (Informatics) and the Parts of Each That are Engineering-Oriented

Until recently it was not unusual for the same thinker to concern himself with philosophy, formal theory, empirical questions, and design problems. So far it would seem that informatics, unlike most other academic disciplines, requires that three types of thinking be acquired in its educational curriculum, as suggested by the Association for Computing Machinery (ACM) and the Institute of Electrical and Electronics Engineers (IEEE): (1) formal theoretic, (2) empirical and experimental, and (3) design-oriented types of thinking (roughly, syntactic, semantic, and pragmatic). A recent discussion (Denning et al., 1989) listed nine sub-areas with these three types of thinking in each (Figure 1); the authors say, "It is the explicit and intricate intertwining of the ancient threads of calculation

	Theory	Abstraction	Design
1. Algorithms and Data Structures			
2. Programming Languages			
3. Architecture			
4. Numerical and Symbolic Computation			
5. Operating Systems			
6. Software Methodology and Engineering			
7. Databases and Information Retrieval			
8. Artificial Intelligence and Robotics			
9. Human - Computer Communication			

SOURCE: Peter J. Denning, et al. 1989. Computing As a Discipline. *Communications of the ACM* 32(January):16-23.

Figure 1. Definition matrix for the computing discipline.

and logical symbol manipulation, together with the modern threads of electronics and electronic representation of information, that gave birth to the discipline of computing." Each of these nine areas has "substantial design and implementation issues" in the purview of engineering. All engineering, like all professions that transform knowledge into action, was always pragmatic, both in the popular and in the technical sense (in linguistics, "pragmatics" is concerned with resolution of ambiguity by recognition of context and appropriate interpretation). Furthermore, adaptability to human beings has always been an engineering consideration, in fact the original motivation, whence some areas have made a special study of it—e.g., systems engineering, industrial engineering, and ergonomics (i.e., man-machine interaction).

But it was always the relationships of material objects and surroundings to human beings that was involved; what about the design of procedures, of general and special purpose languages, of automatic scheduling of distributed devices (including computers) for a distributed audience, of editing systems, of types of communications, of types of graphics, of human sensory reactions to types of prosthetic sensing, and the like? Surely aeronautical and space engineering have been concerned with the last of these. But have we designed types of thinking before and prepared the world at large to live with them?

Just as mathematics has changed into (almost) a pure "knowledge-oriented" discipline, might not computer science do likewise? Because so much of it remains action-oriented, namely programming and the fact that it is equivalent in a way to machine design, must remain a professional type of discipline, just like engineering, which it overlaps.

Moreover it will have the "solved problem syndrome" to an even greater extent than mathematics did (with applied mathematics), because there are so many disciplines involved with symbol manipulation; there may be very many mathematical needs in other disciplines, but there are even more symbolic needs (e.g., diagrammatic plans in sports). Even in mathematics, methodological discussions concerning clumsiness or neatness of notations are examples of applied informatics! And yet the knowledge-oriented and action-oriented parts of computer and information science will still have to remain balanced because the programming activity and machine design activity must, due to their "logical" equivalence.

Adaptation Necessary for Engineers To Participate Fully in Computer Science

Professional disciplines may be expected to change faster than those that are almost purely knowledge-oriented, or those almost purely action-oriented. Because of the professional change syndrome and the solved-problem syndrome the variety of types of engineering of the past and present is due to the variety of knowledge-oriented and action-oriented disciplines they bridge.

Essentially all the engineering types have had their share of applied mathematics, both in their theory and in their computational practice. Most of these theories and computations have been from analysis and from statistics, and this will naturally continue in computer science. All engineers will continue to need numerical analysis in their core education, introductory computer courses leading to computer-aided design (CAD), and probably enough graphics for both displays and to replace, for example, mechanical drawing. However, in addition to these core requirements in their early engineering education, they must still pursue continuing education to keep up with the rapidly changing computer field (the professional change syndrome again). Because of the solved-problem syndrome, they must be prepared either to change over to a resulting new discipline or to switch to new problems.

For computer engineers in particular, their core requirements in mathematics must be more than the analysis courses all engineers need. Abstract algebra and logic are necessary in the understanding of programming language design and capabilities as well as the understanding of the logical limits of computation, where analysis of computational complexity would still not show what is impossible. Since, like all engineers, they must concern themselves with man-machine interaction, some ergonomics should, of course, be part of their education; but because of the psycho-linguistic effects they must be concerned with, they should go deeper into cognitive science. After all, when they design a hardware-software-peopleware system, because of the logical equivalence of hardware and software, they must decide on which parts are more efficient in hardware and which in software; but they must also be concerned with psychological interaction as well. Not only should they be concerned with reasonable interaction between the user and the rest of the system, but with the distribution of users, advisers, machines, and interactive software in extended networks.

References

Bell, E. T. 1937. *Men of Mathematics*. Mineola, N.Y.: Dover.

Boyer, C. B. 1968. *A History of Mathematics*. New York: John Wiley & Sons.

Denning, P. J., D. E. Comer, D. Gries, M. C. Mulder, A. Tucker, A. J. Turner and P. R. Young. 1989. Computing As a Discipline. *Communications of the ACM* 32(January):16-23.

Goldstine, H. H. 1972. *The Computer from Pascal to von Neumann*. Princeton, N.J.: Princeton University Press.

Harmon, M. 1975. *Stretching Man's Mind*. New York: Mason/Charter.

Hofstadter, D. R. 1975. *Gödel, Escher, Bach: An Eternal Golden Braid*. N.Y.: Basic Books.

Kuhn, T. S. 1970. *The Structure of Scientific Revolutions* (2nd enlarged ed.). Chicago: University of Chicago Press.

Luce, R. D., R. R. Bush, and E. Galanter (eds.). 1963. *Handbook of Mathematical Psychology*. New York: John Wiley & Sons.

Luce, R. D., R. R. Bush, and E. Galanter (eds.). 1963. *Readings in Mathematical Psychology*. New York: John Wiley & Sons.

Machlup, F., and U. Mansfield (eds.). 1982. *The Study of Information: Interdisciplinary Messages*. New York: John Wiley & Sons.

McNaughton, R. 1982. *Elementary Computability, Formal Languages, and Automata*. Englewood Cliffs, N.J.: Prentice-Hall.

Metropolis, N., J. Howlett, and G. C. Rota (eds.). 1980. *A History of Computation in the Twentieth Century*. New York: Academic Press.

Penzias, A. 1989. *Ideas and Information—Managing in a High-Tech World*. New York: W. W. Norton.

Pullam, J. M. 1968. *The History of the Abacus*. London: Hutchinson & Company.

Ralston, A. 1971. *Introduction to Programming & Computer Science*. New York: McGraw-Hill.

Randell, B. (ed.). 1975. *The Origins of Digital Computers; Selected Paper*. New York: Springer-Verlag.

Shelley, G. B., and T. J. Cashman. 1984. *Computer Fundamentals for an Information Age*. Brea, CA: Anaheim Publishing Company.

Wood, D. 1987. *Theory of Computation*. New York: Harper & Row.

APPENDIX A
COMPUTING AS A DISCIPLINE

The final report of the Task Force on the Core of Computer Science presents a new intellectual framework for the discipline of computing and a new basis for computing curricula. This report has been endorsed and approved for release by the ACM Education Board.

Peter J. Denning (Chairman), Douglas E. Comer, David Gries,
Michael C. Mulder, Allen Tucker, A. Joe Turner, and Paul R. Young

APPENDIX
A DEFINITION OF COMPUTING AS A DISCIPLINE

Computer science and engineering is the systematic study of algorithmic processes —their theory, analysis, design, efficiency, implementation, and application - that describe and transform information. The fundamental question underlying all of computing is, What can be (efficiently) automated [2.3]. This discipline was born in the early 1940s with the joining together of algorithm theory, mathematical logic, and the invention of the stored-program electronic computer.

The roots of computing extends deeply into mathematics and engineering. Mathematics imparts analysis to the field: engineering imparts design. The discipline embraces its own theory, experimental method, and engineering, in contrast with most physical sciences, which are separate from the engineering disciplines that apply their findings (e.g., chemistry and chemical engineering principles). The science and engineering are inseparable because of the fundamental interplay between the scientific and engineering paradigms within the discipline.

For several thousand years, calculation has been a principal concern of mathematics. Many models of physical phenomena have been used to derive equations whose solutions yield predictions of those phenomena - for example, calculations of orbital trajectories, weather forecasts, and fluid flows. Many general methods for solving such equations have been devised - for example, algorithms for systems of linear equations, differential equations, and integrating functions. For almost the same period, calculations that aid in the design of mechanical systems have been a principal concern of engineering. Examples include algorithms for evaluating stresses in static objects, calculating moments of moving objects, and measuring distances much larger or smaller than our immediate perception.

One product of the long interaction between engineering and mathematics has been mechanical aids for calculating. Some surveyors' and navigators' instruments date back a thousand years. Pascal and Leibniz built arithmetic calculators in the middle 1600s. In the 1930s Babbage conceived of an "analytical engine" that could mechanically and without error evaluate logarithms, trigonometric functions and other general arithmetic functions. His machine, never completed, served as an inspiration for later work. In the 1920, Bush constructed an electronic analog computer for solving general systems of differential equations. In the same period, electromechanical calculating machines capable of addition, subtraction, multiplication, division, and square root computation became available. The electronic flip-flop provided a natural bridge from these machines to digital versions with no moving parts.

Logic is a branch of mathematics concerned with criteria of validity of inference and formal principles of reasoning. Since the days of Euclid, it has been a tool for rigorous mathematical and scientific argument. In the 19th century a search began for a universal system of logic that would be free of the incompletenesses observed in known deductive systems. In a complete system, it would be possible to determine mechanically whether any given statement is either true or false. In 1931, Gödel published his "incompleteness theorem," showing that there is no such system. In the late 1930s, Turing explored the idea of a universal computer that could simulate any step-by-step procedure of any other computing machine. His findings were similar to Gödel's: some well-defined problems cannot be solved by any mechanical procedure. Logic is important not only because of its deep insight into the limits of automatic calculation, but also because of its insight that strings of symbols, perhaps encoded s numbers, can be interpreted both as data and as programs.

This insight is the key idea that distinguishes the stored program computer from calculating machines. The steps of the algorithm are encoded in a machine representation and stored in the memory for later decoding and execution by the processor. The machine code can be derived mechanically from a higher-level symbolic.form, the programming language.

It is the explicit and intricate intertwining of the ancient threads of calculation and logical symbol manipulation, together with the modern threads of electronics and electronic representation of information that gave birth to the discipline of computing.

We identified nine subareas of computing:

1. Algorithms and data structures
2. Programming languages
3. Architecture
4. Numerical and symbolic computation
5. Operating systems
8. Software methodology and engineering
7. Databases and information retrieval
8. Artificial intelligence and robotics
9. Human-Computer communication

Each has an underlying unity of subject matter, a substantial theoretical component, significant abstractions, and substantial design and implementation issues. Theory deals with the underlying mathematical development of the subarea and includes supporting theory such as graph theory, combinatorics, or formal languages. Abstraction (or modelling) deals with models of potential implementations; the models suppress detail, while retaining essential features, and provide means for predicting future behavior. Design deals with he process of specifying a problem, deriving requirements and specifications, iterating and testing prototypes, and implementing a system. Design includes the experimental method, which in computing comes in several styles: measuring programs and systems, validating hypotheses, and prototyping to extend abstractions to practice.

Although software methodology is essentially concerned with design, it also contains substantial elements of theory and abstraction. For this reason, we have identified it as a subarea. On the other hand, parallel and distributed computation are issues that pervades all the subareas and all their components (theory, abstraction, and design); they have been identified neither as subareas nor as subareas components.

The subsequent numbered sections provide the details of each subareas in three parts---theory, abstraction, and design. The boundaries between theory and abstraction, and between abstraction and design are necessarily fuzzy; it is a matter of personal taste where some of the items go.

Our intention is to provide a guide to the discipline by showing its main features, not a detailed map. It is important to remember that this guide to the discipline is not a plan for a course or a curriculum; it is merely a framework in which a curriculum can be designed. It is also important to remember that this guide to the discipline is a snapshot of an organism undergoing constant change. It will require reevaluation and revision at regular intervals.

1. ALGORITHMS AND DATA STRUCTURES

This area deals with specific classes of problems and their efficient solutions. Fundamental questions include: For given classes of problems and their efficient solutions. Fundamental questions include: For given classes of problems, what are the best algorithms: How much storage and time do they require? What is the tradeoff between space and time? What is the worst case of the best algorithms? How well do algorithms behave on average? How general are algorithms—i.e., what classes of problems can be dealt with by similar methods?

1.1 Theory

Major elements of theory in the area of algorithms and data structures are:

1. Computability theory, which defines what machines can and cannot do.
2. Computational complexity theory, which tells how to measure the time and space requirements of computable functions and relates a problem's size with the best- or worst-case performance of algorithms that solve that problem, and provides methods for proving lower bounds on any possible solution to a problem.
3. Time and space bounds for algorithms and classes of algorithms.
4. Levels of intractability: for example, classes of problems solvable deterministically in polynomially bounded time (P-problems); those solvable nondeterministically in

polynomially bounded time (NP-problems); and those solvable nondeterministically in polynomially bounded time (NP-problems); and those solvable efficiently by parallel machines (NC-problems).
5. Parallel computation, lower bounds, and mappings from dataflow requirements of algorithms into communication paths of machines.
6. Probabilistic algorithms, which give results correct with sufficiently high probabilities much more efficiently (in time and space) than determinate algorithms that guarantee their results. Monte Carlo methods.
7. Cryptography.
8. The supporting areas of graph theory, recursive functions, recurrence relations, combinatorics, calculus, induction, predicate and temporal logic, semantics, probability, and statistic.

1.2 Abstraction

Major elements of abstraction in the area of algorithms and data structures are:

1. Efficient, optimal algorithms for important classes of problems and analyses for best, worst and average performance.
2. Classifications of the effects of control and data structure on time and space requirements for various classes of problems.
3. Important classes of techniques such as divide-and-conquer, Greedy algorithms, dynamic programming, finite state machine interpreters, and stack machine interpreters.
4. Parallel and distributed algorithms; methods of partitioning problems into tasks that can be executed in separate processors.

1.3 Design

Major elements of design and experimentation in the area of algorithms and data structures are:

1. Selection, implementation, and testing of algorithms for important classes of problems such as searching, sorting, random-number generation, and textual pattern matching.
2. Implementation and testing of general methods applicable across many classes of problems, such as hashing, graphs, and trees.
3. Implementation and testing of distributed algorithms such as network protocols, distributed data updates, semaphores, deadlock detectors and synchronization methods.
4. Implementation and testing of storage managers such as garbage collection, buddy system, lists, tables, and paging.
5. Extensive experimental testing of heuristic algorithms for combinatorial problems.
6. Cryptographic protocols that permit secure authentication and secret communication.

2. PROGRAMMING LANGUAGES

This area deals with notations for virtual machines that execute algorithms, with notations for algorithms and data, and with efficient translations from high-level languages into machine codes. Fundamental questions include: What are possible organizations of the virtual machine presented by the language (data types, operations, control structures, mechanisms for introducing new types and operations)? How are these abstractions implemented on computers? What notation (syntax) can be used effectively and efficiently to specify what the computer should do?

2.1 Theory

Major elements of theory in the area of programming languages are:

1. Formal languages and automata, including theories of parsing and language translation.
2. Turing machines (base for procedural languages), Post Systems (base for string processing languages), λ-calculus (base for functional languages).
3. Formal semantics: methods for defining mathematical models of computers and the relationships among the models, language syntax, and implementation. Primary methods include denotational, algebraic, operational and axiomatic semantics.
4. As supporting areas: predicate logic, temporal logic, modern algebra and mathematical induction.

2.2 Abstraction

Major elements of abstraction in the area of programming languages include:

1. Classification of languages based on their syntactic and dynamic semantic models; e.g., static typing, dynamic typing, functional, procedural, object-oriented, logic, specification, message passing, and dataflow.
2. Classification of languages according to intended application area; e.g., business data processing, simulation, list processing, and graphics.

3. Classification of major syntactic and semantic models for program structure; e.g., procedure hierarchies, functional composition, abstract data types, and communicating parallel processes.
4. Abstract implementation models for each major type of language.
5. Methods for parsing, compiling, interpretation, and code optimization.
6. Methods for automatic generation of parsers, scanners, compiler components, and compilers.

2.3 Design

Major elements of design and experimentation in the area of programming languages are:

1. Specific languages that bring together a particular abstract machine (semantics) and syntax to form a coherent implementable whole. Examples: procedural (COBOL, FORTRAN, ALGOL, Pascal, Ada, C), functional (LISP), dataflow (SISAL, VAL), object-oriented (Smalltalk, CLU), logic (Prolog), strings (SNOBOL), and concurrency (CSP, Occam, Concurrent Pascal, Modula 2).
2. Specific implementation methods for particular classes of languages: run-time models, static and dynamic execution methods, typing checking, storage and register allocation, compilers, cross compilers, and interpreters, systems for finding parallelism in programs.
3. Programming environments.
4. Parser and scanner generators (e.g., YACC, LEX), compiler generators.
5. Programs for syntactic and semantic error checking, profiling, debugging, and tracing.
6. Applications of programming-language methods to document-processing functions such as creating tables, graphs, chemical formulas, spreadsheets equations, input and output, and data handling. Other applications such as statistical processing.

3. ARCHITECTURE

This area deals with methods of organizing hardware (and associated software) into efficient, reliable systems. Fundamental questions include: What are good methods of implementing processors, memory, and communication in a machine? How do we design and control large computational systems and convincingly demonstrate that they work as intended despite errors and failures? What types of architectures can efficiently incorporate many processing elements that can work concurrently on a computation? How do we measure performance?

3.1 Theory

Major elements of theory in the area of architecture are:

1. Boolean algebra.
2. Switching theory.
3. Coding theory.
4. Finite state machine theory.
5. The supporting areas of statistics, probability, queueing, reliability theory, discrete mathematics, number theory, and arithmetic in different number systems.

3.2 Abstraction

Major elements of abstraction in the area of architecture are:

1. Finite state machine and Boolean algebraic models of circuits that relate function to behavior.
2. Other general methods of synthesizing systems from basic components.
3. Models of circuits and finite state machines for computing arithmetic functions over finite fields.
4. Models for data path and control structures.
5. Optimizing instruction sets for various models and workloads.
6. Hardware reliability: redundancy, error detection, recovery, and testing.
7. Space, time, and organizational tradeoffs in the design of VLSI devices.
8. Organization of machines for various computational models: sequential, dataflow, list processing, array processing, vector processing, and message-passing.
9. Identification of design levels; e.g., configuration, program, instruction set, register, and gate.

3.3 Design

Major elements of design and experimentation in the area of architecture are:

1. Hardware units for fast computation; e.g., arithmetic function units, cache.
2. The so-called von Neumann machine (the single-instruction sequence stored program computer); RISC and CISC implementations.
3. Efficient methods of storing and recording information, and detecting and correcting errors.

4. Specific approaches to responding to errors; recovery, diagnostics, reconfiguration, and backup procedures.
5. Computer aided design (CAD) systems and logic simulations for the design of VLSI circuits. Production programs for layout, fault diagnosis. Silicon compilers.
6. Implementing machines in various computational models; e.g., dataflow, tree, LISP, hypercube, vector, and multiprocessor.
7. Supercomputers, such as the Cray and Cyber machines.

4. NUMERICAL AND SYMBOLIC COMPUTATION

This area deals with general methods of efficiently and accurately solving equations resulting from mathematical models of systems. Fundamental questions include: How can we accurately approximate continuous or infinite processes by finite discrete processes? How do we cope with the errors arising from these approximations? How rapidly can a given class of equations be solved for a given level of accuracy? How can symbolic manipulations on equations, such as integration, differentiation, and reduction to minimal terms, be carried out? How can the answers to these questions be incorporated into efficient, reliable, high-quality mathematical software packages?

4.1 Theory

Major elements of theory in the area of numerical and symbolic computation are:

1. Number theory.
2. Linear algebra.
3. Numerical analysis.
4. Nonlinear dynamics.
5. The supporting areas of calculus, real analysis, complex analysis, and algebra.

4.2 Abstraction

Major elements of abstraction in the area of numerical and symbolic computation are:

1. Formulations of physical problems as models in continuous (and sometimes discrete) mathematics.
2. Discrete approximations to continuous problems. In this context, backward error analysis, error propagation and stability in the solution of linear and nonlinear systems. Special methods in special cases, such as Fast Fourier Transform and Poisson solvers.
3. The finite element model for a large class of problems specifiable by regular meshes and boundary values. Associated iterative methods and convergence theory; direct, implicit, multigrids, rates of convergence. Parallel solution methods. Automatic grid refinement during numerical integration.
4. Symbolic integration and differentiation.

4.3 Design

Major elements of design and experimentation in the area of numerical and symbolic computation are:

1. High-level problem formulation systems such as CHEM and WEB.
2. Specific libraries and packages for linear algebra, ordinary differential equations, statistics, nonlinear equations, and optimizations; e.g., LINPACK, EISPEACK, ELLPACK.
3. Methods of mapping finite element algorithms to specific architectures—e.g., multigrids on hypercubes.
4. Symbolic manipulators, such as MACSYMA and REDUCE, capable of powerful and nonobvious manipulations, notably differentiations, integrations, and reductions of expressions to minimal terms.

5. OPERATING SYSTEMS

This area deals with control mechanisms that allow multiple resources to be efficiently coordinated in the execution of programs. Fundamental questions include: What are the visible objects and permissible operations at each level in the operation of a computer system? For each class of resource (objects visible at some level), what is a minimal set of operations that permit their effective use? How can interfaces be organized so that users deal only with abstract versions of resources and not with physical details of hardware? What are effective control strategies for job scheduling, memory management, communication among concurrent tasks, reliability, and security? What are the principles by which systems can be extended in function by repeated application of a small number of construction rules? How should distributed computations be organized so that many autonomous machines connected by a communication network can participate in a computation, with the details of network protocols, host locations, band-widths, and resource naming being mostly invisible?

5.1 Theory

Major elements of theory in the area of operating systems are:

1. Concurrency theory; synchronization, determinacy, and deadlocks.
2. Scheduling theory, especially processor scheduling.
3. Program behavior and memory management theory, including optimal policies for storage allocation.
4. Performance modeling and analysis.
5. The supporting areas of bin packing, probability, queueing theory, queueing networks, communication and information theory, temporal logic, and cryptography.

5.2 Abstraction

Major elements of abstraction in the area of operating systems are:

1. Abstraction principles that permit users to operate on idealized versions of resources without concern for physical details (e.g., process rather than processor, virtual memory rather than main-secondary hierarchy, files rather than disks).
2. Binding of objects perceived at the user interface to internal computational structures.
3. Models for important subproblems such as process management, memory management, job scheduling, secondary storage management, and performance analysis.
4. Models for distributed computation; e.g., clients and servers, cooperating sequential processes, message-passing, and remote procedure calls.
5. Models for secure computing; e.g., access controls, authentication, and communication.
6. Networking, including layered protocols, naming, remote resource usage, help services, and local network protocols such as token-passing and shared buses.

5.3 Design

Major elements of design and experimentation in the area of operating systems are:

1. Prototypes of time sharing systems, automatic storage allocators, multilevel schedulers, memory managers, hierarchical file systems and other important system components that have served as bases for commercial systems.
2. Techniques for building operating systems such as UNIX, Multics, Mach, VMS, and MS-DOS.
3. Techniques for building libraries of utilities; e.g., editors, document formatters, compilers, linkers, and device drivers.
4. Files and file systems.
5. Queueing network modeling and simulation packages to evaluate performance of real systems.
6. Network architectures such as ethernet, FDDI, token ring nets, SNA, and DECNET.
7. Protocol techniques embodied in the Department of Defense protocol suite (TCP/1P), virtual circuit protocols, internet, real time conferencing, and X.25.

6. SOFTWARE METHODOLOGY AND ENGINEERING

This area deals with the design of programs and large software systems that meet specifications and are safe, secure, reliable, and dependable. Fundamental questions include: What are the principles behind the development of programs and programming systems? How does one prove that a program or system meets its specifications? How does one develop specifications that do not omit important cases and can be analyzed for safety? How do software systems evolve through different generations? How can software be designed for understandability and modifiability?

6.1 Theory

Major elements of theory in the area of software methodology and tools are:

1. Program verification and proof.
2. Temporal logic.
3. Reliability theory.
4. The supporting areas of predicate calculus, axiomatic semantics, and cognitive psychology.

6.2 Abstraction

Major elements of abstraction in the area of software methodology and tools are:

1. Specification methods, such as predicate transformers, programming calculi, abstract data types, and Floyd-Hoare axiomatic notations.
2. Methodologies such as stepwise refinement, modular design, modules, separate compilation, information-hiding, dataflow, and layers of abstraction.

3. Methods for automating program development; e.g., text editors, syntax-directed editors, and screen editors.
4. Methodologies for dependable computing; e.g., fault tolerance, security, reliability, recovery, *N*-version programming, multiple-way redundancy, and check-pointing.
5. Software tools and programming environments.
6. Measurement and evaluation of programs and systems.
7. Matching problem domains through software systems to particular machine architectures.
8. Life cycle models of software projects.

6.3 Design

Major elements of design and experimentation in the area of software methodology and tools are:

1. Specification languages (e.g., PSL 2, IMA JO), configuration management systems (e.g., in Ada APSE), and revision control systems (e.g., RCS, SCCS).
2. Syntax directed editors, line editors, screen editors, and word processing systems.
3. Specific methodologies advocated and used in practice for software development; e.g., HDM and those advocated by Dijkstra, Jackson, Mills, or Yourdon.
4. Procedures and practices for testing (e.g., walk-through, hand simulation, checking of interfaces between modules, program path enumerations for test sets, and event tracing), quality assurance, and project management.
5. Software tools for program development and debugging, profiling, text formatting, and database manipulation.
6. Specification of criteria levels and validation procedures for secure computing systems, e.g., Department of Defense.
7. Design of user interfaces.
8. Methods for designing very large systems that are reliable, fault tolerant, and dependable.

7. DATABASE AND INFORMATION RETRIEVAL SYSTEMS

This area deals with the organization of large sets of persistent, shared data for efficient query and update. Fundamental questions include: What modeling concepts should be used to represent data elements and their relationships? How can basic operations such as store, locate, match, and retrieve be combined into effective transactions? How can these transactions interact effectively with the user? How can high-level queries be translated into high-performance programs? What machine architectures lead to efficient retrieval and update? How can data be protected against unauthorized access, disclosure, or destruction? How can large databases be protected from inconsistencies due to simultaneous update? How can protection and performance be achieved when the data are distributed among many machines? How can text be indexed and classified for efficient retrieval?

7.1 Theory

Major elements of theory in the area of databases and information retrieval systems are:

1. Relational algebra and relational calculus.
2. Dependency theory.
3. Concurrency theory, especially serializable transactions, deadlocks, and synchronized updates of multiple copies.
4. Statistical inference.
5. Sorting and searching.
6. Performance analysis.
7. As supporting theory; cryptography.

7.2 Abstraction

Major elements of abstraction in the area of databases and information retrieval systems are:

1. Models for representing the logical structure of data and relations among the data elements, including the relational and entity-relationship models.
2. Representations of files for fast retrieval, such as indexes, trees, inversions, and associative stores.
3. Methods for assuring integrity (consistency) of the data base under updates, including concurrent updates of multiple copies.
4. Methods for preventing unauthorized disclosure or alteration and for minimizing statistical inference.
5. Languages for posing queries over databases of different kinds (e.g., hypertext, text, spatial, pictures, images, rule-sets). Similarly for information retrieval systems.
6. Models, such as hypertext, which allow documents to contain text at multiple levels and to include video, graphics, and voice.
7. Human factors and interface issues.

7.3 Design

Major elements of design in the area of database and information retrieval systems are:

1. Techniques for designing databases for relational, hierarchical, network, and distributed implementations.
2. Techniques for designing database systems such as INGRES, System R, dBase III, and DB-2.
3. Techniques for designing information retrieval systems such as LEXIS, Osiris, and Medline.
4. Design of secure database systems.
5. Hypertext systems such as NLS, NoteCards, Intermedia, and Xanadu.
6. Techniques to map large databases to magnetic disk stores.
7. Techniques for mapping large, read-only databases onto optical storage media—e.g., CD/ROM and WORMS.

8. ARTIFICIAL INTELLIGENCE AND ROBOTICS

This area deals with the modeling of animal and human (intelligent) behavior. Fundamental questions include: What are basic models of behavior and how do we build machines that simulate them? To what extent is intelligence described by rule evaluation, inference, deduction, and pattern computation? What is the ultimate performance of machines that simulate behavior by these methods? How are sensory data encoded so that similar patterns have similar codes? How are motor codes associated with sensory codes? What are architectures for learning systems, and how do those systems represent their knowledge of the world?

8.1 Theory

Major elements of theory in the area of artificial intelligence and robotics are:

1. Logic, e.g., monotonic, nonmonotonic, and fuzzy.
2. Conceptual dependency.
3. Cognition.
4. Syntactic and semantic models for natural language understanding.
5. Kinematics and dynamics of robot motion and world models used by robots.
6. The supporting areas of structural mechanics, graph theory, formal grammars, linguistics, philosophy, and psychology.

8.2 Abstraction

Major elements of abstraction in the area of artificial intelligence and robotics are:

1. Knowledge representation (e.g., rules, frames, logic) and methods of processing them (e.g., deduction, inference).
2. Models of natural language understanding and natural language representations, including phoneme representations; machine translation.
3. Speech recognition and syntheses, translation of text to speech.
4. Reasoning and learning models; e.g., uncertainty, nonmonotonic logic, Bayesian inference, beliefs.
5. Heuristic search methods, branch and bound, control search.
6. Machine architectures that imitate biological systems, e.g., neural networks, connectionism, sparse distributed memory.
7. Models of human memory, autonomous learning, and other elements of robot systems.

8.3 Design

Major elements of design and experimentation in artificial intelligence and robotics include:

1. Techniques for designing software systems for logic programming, theorem proving, and rule evaluation.
2. Techniques for expert systems in narrow domains (e.g., Mycin, Xcon) and expert system shells that can be programmed for new domains.
3. Implementations of logic programming (e.g., PROLOG).
4. Natural language understanding systems (e.g., Margle, SHRDLU, and preference semantics).
5. Implementations of neural networks and sparse distributed memories.
6. Programs that play checkers, chess, and other games of strategy.
7. Working speech synthesizers, recognizers.
8. Working robotic machines, static and mobile.

9. HUMAN-COMPUTER COMMUNICATION

This area deals with the efficient transfer of information between humans and machines via various human-like sensors and motors, and with information structures that reflect human conceptualizations. Fundamental questions include: What are efficient methods of representing objects and automatically creating pictures for viewing? What are effective methods for receiving input or presenting output? How can the risk of misperception and subsequent human error be minimized? How can graphics and other tools be used to understand physical phenomena through information stored in data sets?

9.1 Theory

Major elements of theory in human-computer communication are:

1. Geometry of two and higher dimensions including analytic, projective, affine, and computational geometrics.
2. Color theory.
3. Cognitive psychology.
4. The supporting areas of Fourier analysis, linear algebra, graph theory, automats, physics, and analysis.

9.2 Abstraction

Major elements of abstraction in the area of human-computer communication are:

1. Algorithms for displaying pictures including methods for smoothing, shading, hidden lines, ray tracing, hidden surfaces, transparent surfaces, shadows, lighting, edges, color maps, representations by splines, rendering, texturing, antialiasing, coherence, fractals, animation, representing pictures as hierarchies of objects.
2. Models for computer-sided design (CAD).
3. Computer representations of physical objects.
4. Image processing and enhancement methods.
5. Man-machine communication, including psychological studies of modes of interaction that reduce human error and increase human productivity.

9.3 Design

Major elements of design and experimentation in the area of human-computer communication are:

1. Implementation of graphics algorithms on various graphics devices, including vector and rastor displays and a range of hardcopy devices.
2. Design and implementation of experimental graphics algorithms for a growing range of models and phenomena.
3. Proper use of color graphics for displays; accurate reproduction of colors on displays and hardcopy devices.
4. Graphics standards (e.g., GKS, PHIGS, VDI), graphics languages (e.g., PostScript), and special graphics packages (e.g., MOGLI for chemistry).
5. Implementation of various user interface techniques including direct manipulation on bitmapped devices and screen techniques for character devices.
6. Implementation of various standard file interchange formats for information transfer between differing systems and machines.
7. Working CAD systems.
8. Working image enhancement systems (e.g., at JPL for pictures received from space probes).

ACKNOWLEDGMENTS

Many people generously provided written comments in response to drafts of this report. Although it was not possible to accommodate every comment in detail, we did take every comment into account in revising this report. We are grateful to the following people for sending us their comments:

Paul Abrahams
J. Mack Adams
Robert Aiken
Donald Bagert
Alan Blomann
Frank Bonach
Richard Botting
Albert Briggs, Jr.
Judy Brown
Rick Carlson
Thomas Cheatham
Neal Coulter
Steve Cunningham
Verlynda Dobbs
Caroline Eastman
Richard Epstein
Frank Friedman
C. W. Gear
Robert Glass
Nico Habermann
Judy Hankins
Charles Keleman
Ken Kennedy
Elliot Koffman
Barry Kurtz
Doris Lidke
Michael Loui
Paul Laker
Susan Merritt
John Moti
J. Paul Myers
Bob Noonan
Alan Portis
Jesse Poore
Terrence Pratt
Jean Rogers
Jean Sammet
Mary Shaw
J. W. Smith
Dennis Smolarski
Ed Upchurch
Garret White
Gio Wiederfield
Stuart Zweben

References

1. Abelson, M., and Sussman, G. *Structure and Interpretation of Computer Programs.* MIT Press, Cambridge, Mass., 1983
2. Arden, B., ed. *See What Can Be Automated?* Report of the NSF Computer Science and Engineering Research Study (COSERS). MIT Press. Cambridge, Mass., 1980.
3. Denning, P. What is computer science? *Am. Sci.* 73 (Jan.-Feb. 1985). 16-19.
4. Flores, F., and Graves, M. Education. (working paper available from Logonet, Inc. 2200 Powell Street, 11th Floor, Emeryville, Calif. 94608.)
5. Nowell, A., Perlis, A., and Simon, H. What is computer science: *Sci.* 157 (1987), 1373-1374. reprinted in Abacus 4, (Summer 1987). 32.

(copied from *Communications of the ACM*, January 1989 Volume 32, Number 1, pg. 16-23)